Indonesian Meliponiculture & Beyond

Author: Abu Hassan Jalil
Editor: David W. Roubik

NBB

IBRA
INTERNATIONAL BEE
RESEARCH ASSOCIATION

Jointly published by: The International Bee Research Association,
a Company Limited by Guarantee, 1, Agincourt Street, Monmouth, NP25 3DZ (UK) &
Northern Bee Books, Scout Bottom Farm, Mytholmroyd, Hebden Bridge HX7 SJS (UK).

Obtainable from:
www.ibra.org.uk & www.northernbeebooks.co.uk

ISBN: 978-1-913811-10-5

Artwork by DM Design & Print
IBRA Proof Editor - Stuart A. Roberts

Indonesian Meliponiculture & Beyond

Synopsis:

The 3rd instalment in a series of literature on ASEAN Meliponiculture and Beyond, this portion entails the varied and diverse Stingless Bee Fauna in the Indonesian archipelago.

We attempt to bridge the different beekeeping and culture methods in hiving, colony acquisition or multiplication, forage provisioning, nest building material sources and regulating extreme weather or local conditions.

To address conservation across the regional ASEAN divide, we produce this literature in a dual-language format for comprehension by a multi-lingual and multi-ethnic population.

The methodology employed here will vary from the common to more sophisticated innovations, and at times beyond basic product development.

This work also looks at honey-making bee species diversity in the Greater and Lesser Sunda Regions, Wallacea, and the Papuasia Region of the Indonesian Archipelago.

It is a vast expanse of thousands of Islands and varied flora and fauna. We have compiled literary sources from many academicians, enthusiasts and breeders of stingless bees. A directory of each culture's varied regional and vernacular names is also included.

Foreword

Bisismillahir rahmanir Rahim,

We thank God for the presence of Allah SubhanaHu Wa Ta'ala, who has given strength, perseverance and patience to the author so that the book "Indonesian Meliponiculture & Beyond" finally reaches the readers. We also recite sholawat in the name of the Prophet Muhammad Shalulla Hu 'Alaihi Wasallam, who has shown the right path, may sholawat and greetings are bestowed on his family and followers until the end of time.

Based on the author's travel experience around Indonesia, the book "Indonesian Meliponiculture & Beyond" was published. Besides that, the main point of this writing is his deep connection with many researchers and beekeeping practitioners in Indonesia and his efforts to explore the potential of Indonesian flora that support native stingless bees.

This book is specially prepared for those who want knowledge of the techniques of stingless beekeeping because, along with the journey, many experiences are contained in this book. This narrative is especially so that the public can learn about beekeeping activities in Indonesia. It may inspire many, to know the condition of beekeeping, especially in Indonesia, and to try to let this guide helpful in planning useful activities.

This book consists of 6 sections; Section 1 contains Aceh, West Sumatra, Riau and Lampung; Section 2 contains Java, Nusa Kambangan, Madura and Bali; section 3 contains West Kalimantan, South Kalimantan and East Kalimantan; Section 4 contains Lombok Island and East Nusa Tenggara; Section 5 contains Sulawesi, Maluku Islands, West Papua Regency & Papua Indonesia; and the last one contains Miscellaneous papers. In this book, we can see the Main Index and the Appendix that indicate the material of each Section.

The writing of the book "Indonesian Meliponiculture & Beyond" began in 2017 with a record of the author's journey to Indonesia, then the materials were collected, and the writing was improved so that this book was realized.

Through this foreword, of course, the author would like to express his gratitude to various parties who have assisted in the publication of this book. The author also feels that this book is far from perfect. In the next writing or the next edition, it will be better. Therefore, all suggestions, comments and constructive criticism are highly welcome.

New economic growth is based on local resources so that the Indonesian economy will be evenly strong throughout the archipelago, and move forward to build the nation. In addition, cooperation between regions or countries will be established so that mutually supportive and beneficial friendships are formed by all parties. The most important thing seems to provide materials for public welfare and health that allow self-management. Finally, I hope that the publication of the book "Indonesian Meliponiculture & Beyond" will create practitioners in beekeeping, especially stingless bees, which will grow home industries or micro-enterprises, and perhaps even large export-oriented industrial enterprises.

Bogor, January 1, 2022

Dr. Ir. Akhmad Endang Zainal Hasan, M. Si

Lecturer of Biochemistry, Faculty of Mathematics and Natural Sciences

IPB University, Indonesia

Editor's Prologue

Worldwide, led by Australia, SE Asia, Latin America, and Africa, researchers now ask the right questions, use the right tools, and provide intriguing answers on naturally harvested products of social bee colonies in the tropics. Meliponini uses many more than do honey bees. This is expected, because those little stingless bees have held the tropics in their daily activities for over twice as long as the honey bees, and have made many discoveries and applications in their tiny world, as social insects. We are still ignorant of what exactly they do and how they do it. We would have to follow them to the forest canopy and deep underground or in tree hollows to find out much more than they are bees that visit flowers and give honey. But there is much more, and we are beginning to see its importance and know some of the relevant facts.

Here, in a scholarly and deeply human endeavour, AHJ helps us find the source of some of our traditional medicines and the ecology and natural history of the little social bees that make this possible. Curiosity aside, the economics are intriguing. Economics aside, natural history and science cannot fail. Please enjoy this work as readily as I have, and find what it means for each of you.

March 1st, 2022

Dr. David W. Roubik

Senior Scientist

Smithsonian Tropical Research Institute

Republic of Panama

Contents

Preface

After acquiring much knowledge via the travels throughout the maritime ASEAN network and firsthand the diverse bee fauna, I was inspired to write on the various beekeeping techniques practised by people of many different ethnic backgrounds. Some are traditionally handed down, while others are innovations applied within higher learning and research institutions. The different weather cycles in different islands also created a varied methodology for extreme weather protection, out of necessity.

Sabang, the westernmost tiny island in Aceh, and Merauke, the easternmost town in Irian Jaya, give rise to a general narrative for travel in Indonesia. One can only sigh in awe of the numerous islands in terms of language and culture, let alone the flora and fauna. The anecdote "from Sabang to Merauke" readily describes the expanse of the Indonesian archipelago.

This book will attempt to explain the valid concepts and misguided misconceptions that have been established in everyday activities among regional beekeeping practices. I have split up the subject into six sections, i.e., 1) Sumatra, 2) Kalimantan, 2) Java & Bali, 4) Lombok & Sulawesi, 5) Maluku & Papua, and 6) Miscellaneous relevant papers that help to fill in the picture beyond Meliponiculture.

Section 1 – Aceh, West Sumatra, Riau & Lampung

More often than not, we indulge our curiosity as meliponiculturists as to how others do the same occupation. We also frequently wonder what the peculiarities and diversity of species in other parts of our region are, especially the Indo-Malayan region. Sumatra, one of the greater Sunda Islands, has diverse species and diverse cultures, ethnic groups and dialects. Faced with a rather vast area to cover, we broke our tour in three legs. First is the southernmost district of Lampung for a 2-day tour. Our basic strategy was to minimize road trips and pick towns conveniently with airports. This preference brought us into the 2nd Leg, covering the central districts. As such, we picked Padang in West (Central) Sumatra (aka SUMBAR, i.e., Sumatra Barat). Here the aim was to touch down in Minangkabau to view and observe *Tetragonula minangkabau* (Sakagami and Inoue, 1985) in its natural habitat or domestication. On the 2nd leg from Sumbar, we travelled to Pekanbaru in the Riau Province. Pekanbaru has an airport to continue on our 3rd leg in the North served by the Medan airport. Even this northern region was still quite vast, and we could only cover the East Aceh Districts. This route brought us through from Binjai, Medan right up to the East Aceh/ North Aceh border of Lokhseumawe. Here in east Aceh, we found the term Kelulut for meliponines was now called Linoet (Acehnese term). South to North, we covered 15 sites, and Lokhseumawe is where we ended our tour.

Section 2 – Java, Nusa Kambangan, Madura & Bali.

This portion constitutes a field trip in Bogor, West Java. The Malaysian delegation from a conference in Jakarta went with a representative from Thailand (Dr. Bajaree Chuttong). Included also are some reports of Meliponiculture activities in Central Java. One in Banyumas, Central Java, reported by AHJ and another in Pati, Central Java, reported by Norita Widya Pangestika.

The third portion is a report on the East Java and Bali meliponiculture tour conducted after the conference (although independent of the AAA 2018 conference activity). This portion was reported by Norita in Bahasa Indonesia and translated to English, and edited by AHJ with additions of photo albums from Bajaree. Bajaree also provided the bibliography relevant to this portion. The said tour started in Wonosari and covered the southern routes until Trenggalek, where we shifted up to Tulungagung headed towards Surabaya. We crossed the Madura straits by ferry to arrive at Kamal, Bangkalan in Madura. From here, we travelled back to Surabaya (capital of East Java) to spend a night and next day headed for Probolinggo and later to Bondowoso. After Bondowoso, we travelled to Banyuwangi at the eastern end of Java with the plan to take the ferry to Bali the next morning.

Then, we crossed over to Bali from Banywangi (where we spent a night nearest to the ferry terminal). We travelled the northern coastal route to Kintamani, then to Bangli and finally to Karangasem in Bali. There we spent another night at a Beach Hotel. The next day was a trip south to Sanur, where we ended the tour.

Section 3 – West Kalimantan, South Kalimantan & East Kalimantan.

The thought of exploring Kalimantan was quite daunting, and I knew that it was an uphill task. I allocated ten days and saw how far it would get me. The ten-day plan was what most enthusiasts would want to spend on a Meliponiculture tour. I picked cities that had airports to minimize road trips. Being a Malaysian, I decided to start in Kucing, Sarawak. From Kucing, I travelled to Lundu, which was close to the Malaysian border town of Kg Biawak. I would cross into Kalimantan via Aruk (the Indonesian Border town). I arranged my itinerary with advice from beekeeper groups on social media.

The first task would be a road trip that would take me to Sambas and then on a long drive to Pontianak. After completing Pontianak and the suburban areas (Galang), West Kalimantan, I would take a flight from Pontianak to Ketapang and its suburbs, like Sukadana. After this, there is Pangkalan Bun, where I would transit for a few hours before heading to Sampit in Central Kalimantan. Having only one night to spend in Sampit, I would fly to Banjar Baru the next morning and Barabai in South Kalimantan. My next flight would be from Banjarmasin to Balikpapan in East Kalimantan. Here, in East Kalimantan, my final destination would be Samarinda and Tenggarong and after returning to Balikpapan, head back to Kuala Lumpur. This stop ends my 10-day Meliponiculture tour of Kalimantan, where I meet and interact with more than 20 groups of beekeepers throughout the 23 locations in 4 provinces (West, Central, South & East Kalimantan).

When I arrived in Kalimantan, I found apparently in Borneo the language had many dialects and so much variety. I first faced the Kedayan people, especially in the borders of Sarawak & West Kalimantan, then Dayak Melayik in Sambas & the main ethnic Dayak in Pontianak, and Dayak Pesaguan in Ketapang, Dayak Pasir Panjang in Pangkalan Bun. In Sampit, there was also the Dayak Sampit tribe, sub-ethnic Ngaju, after which Pak Marhadiyansyah introduced me to Banjar Language, even though there were Dayak Bukit, ethnic Pitap and Dayak Meratus in Barabai. In Samarinda, I faced peoples of Dayak Pampang, ethnic Kenyah (according to Dr. Syafrizal Fachmi, many carving motifs and lay-out structural designs in Samarinda were based on Kenyah motifs). At the journey's end of Tenggarong, I met with some Beekeepers of the Kutai heritage.

Section 4 -Lombok Island & East Nusa Tenggara.

Nyanteng or Teldan is the vernacular name for the stingless bee in Lombok, West Nusa Tenggara. Lombok is one of the small 'Lesser Sunda' islands in western Indonesia. Its location East of the Wallace Line foretells the island of Lombok would have different flora and fauna existing in Java, Bali, Sumatra & Kalimantan. Suspected on the island of Lombok, there are two species of *Tetragonula* cf. *sapiens* and *Tetragonula* cf. *clypearis* (Riendriasari, 2014). It will be an interesting effort to scrutinize and learn the morphology and behaviour of the stingless bees in Lombok. Another matter of interest is to learn how the Lombok local inhabitants or indigenous community conduct the cultivation of these bees and to gather indigenous knowledge and wisdom of Sasak people in matters of Honey harvesting and utilization.

Interestingly, the stingless bees in Lombok use the Arenga Palm's (*Arenga pinnata*) spent fruit stalk to develop their nests. Comparing this to the occurrences of wild *Tetragonula* spp. nests in Java, none utilize such palm hosts. It was also found that the local community have an aversion to utilizing dried out gourds and Maja fruit as bee hives to reduce costs of constructing wooden hive boxes. They seem more innovative in using locally available materials to construct hive shelves, pedestals and palm thatch roofing.

Section 5 – Sulawesi, Maluku Islands & West Papua Regency and Papua Indonesia.

We next had a look at the meliponiculture of *Wallacetrigona incisa* and bee forage in Sulawesi, Wallacea Region of Indonesia. There appears to occur *Tetragonula* nr. *biroi* and a few other *Tetragonula* spp. in different parts of the large Wallacea Island of Sulawesi. This section is dedicated to the late Suardi Saide, an ardent beekeeper who regularly discussed stingless bee characteristics and foraging habits.

Subsequently, we explore the initiatives and potential of Meliponiculture of Papuasia Meliponines in the Maluku Islands, along with the vegetation and the threats, strengths and weaknesses to beekeeping there. This section also covers the Meliponiculture activities in West Papua Regency and Papua, including the Sahul Shelf peculiarities. The National Park of Wasur is very close to the border between Indonesian Papua and Papua New Guinea. At this park, we explore the types of Papuasian stingless bees with images from the Park Rangers.

Section 6 – Beyond Meliponiculture.

There are in-depth accounts on the standards of Propolis and the antibacterial and anti-cancer properties of Meliponine Propolis. A paper discusses the extra nest structure development after a predator attack. Finally, there is discussion and an elaboration on the ecosystem of Meliponiculture in the Tropics. The idea is to consolidate all factors that constitute the Meliponiculture ecosystem into a bubble for quick reference. It also encompasses different meliponiculture techniques and methods in all the locations on the Major islands throughout the Indonesian Archipelago.

Appendices include names of the many types of wood in Indonesia and vernacular names of Meliponines throughout the Indonesian Archipelago. This vernacular name list shows the diverse language and culture found throughout, with about 55 different names. I also compare the vernacular names of Stingless Bees in The Philippines, Malaysian Borneo and Brunei. The second appendix lists woods' trade names compilations to Indonesia's botanical and vernacular names. It helps to exhibit the array of kinds of wood available for hive construction and or host trees in need of conservation.

0.0 Auxiliary Chapter - Preambles

BIOLOGY AND DIVERSITY OF STINGLESS BEES

(Illustrated Notes and Tabulations)

By Prof. Dr. Siti Salmah

Email: salmah.idrus@gmail.com

0.1 INTRODUCTION

They are some basics to the practice of Meliponiculture in Indonesia, and they contain the necessary initial knowledge required. Stingless bees in Indonesia are known and recognized with many names. In the multi-lingual cultures of Indonesia, the common names used are some of the following:

1. Kelulut (Kalimantan)
2. Galo-galo (Riau-Minang)
3. Lanceng (Jawa)
4. Teuwel (Sunda)
5. Emmu (Sulawesi)

The taxonomic classification is as follows:

Family: Apidae

Subfamily: Apinae

Tribe: Meliponini

Order: : Hymenoptera

We provide here some comparisons of the different bees in the family Apidae

Below is a size comparison chart of some examples of stingless bees with other tribes.

0.2 World Distribution of Stingless Bees

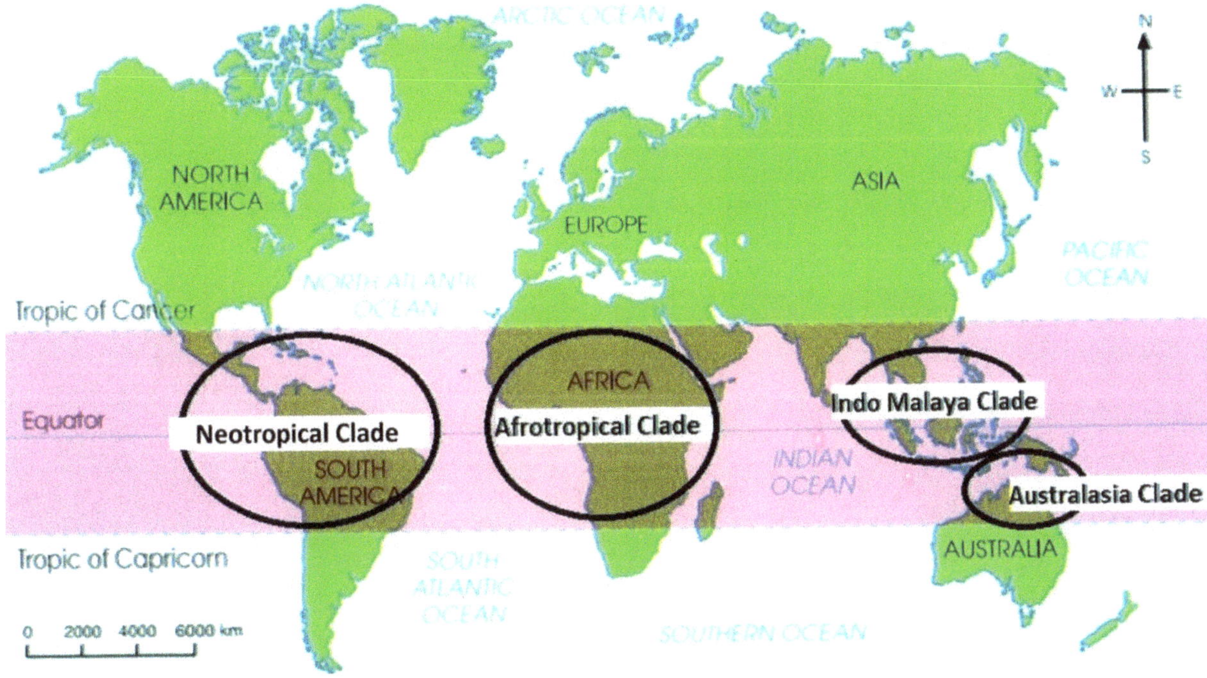

Stingless Bee abundance in various habitats and altitudes occupied by other bee species

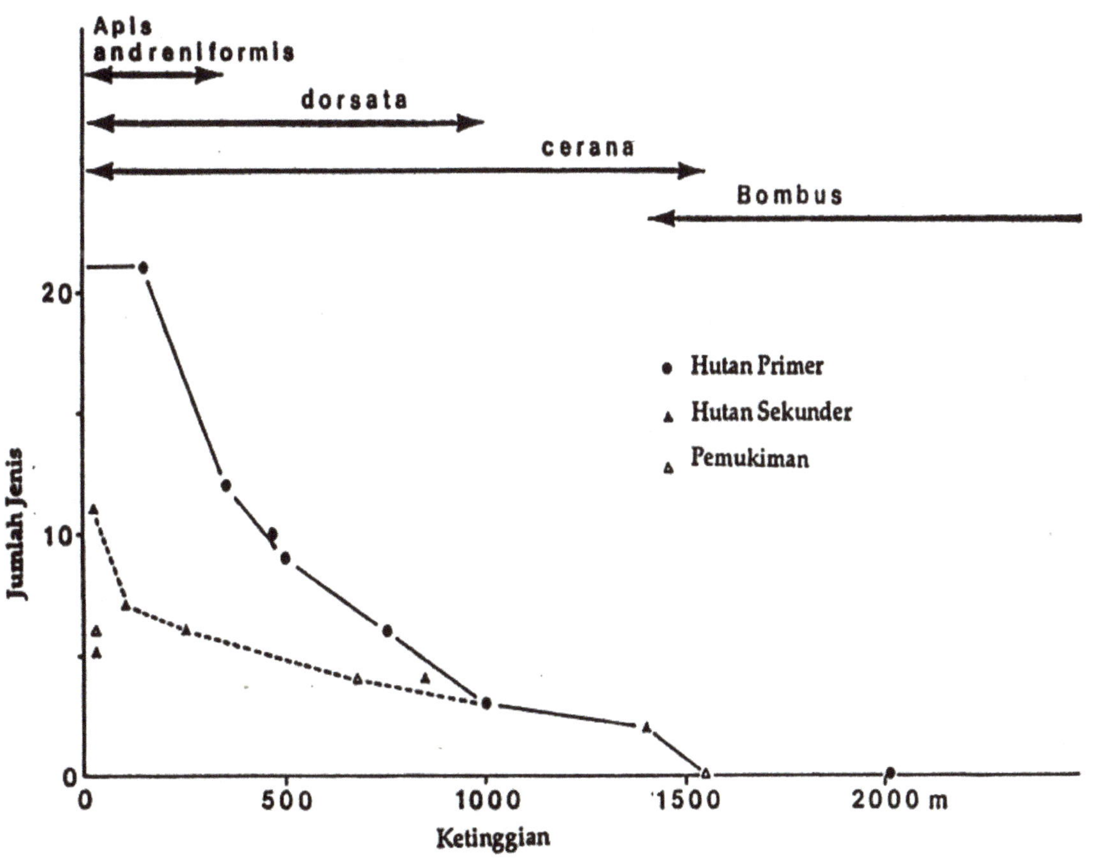

0.3 Morphology

The Morphological Structure of the Stingless Bee (Indonesian terminology of anatomy)

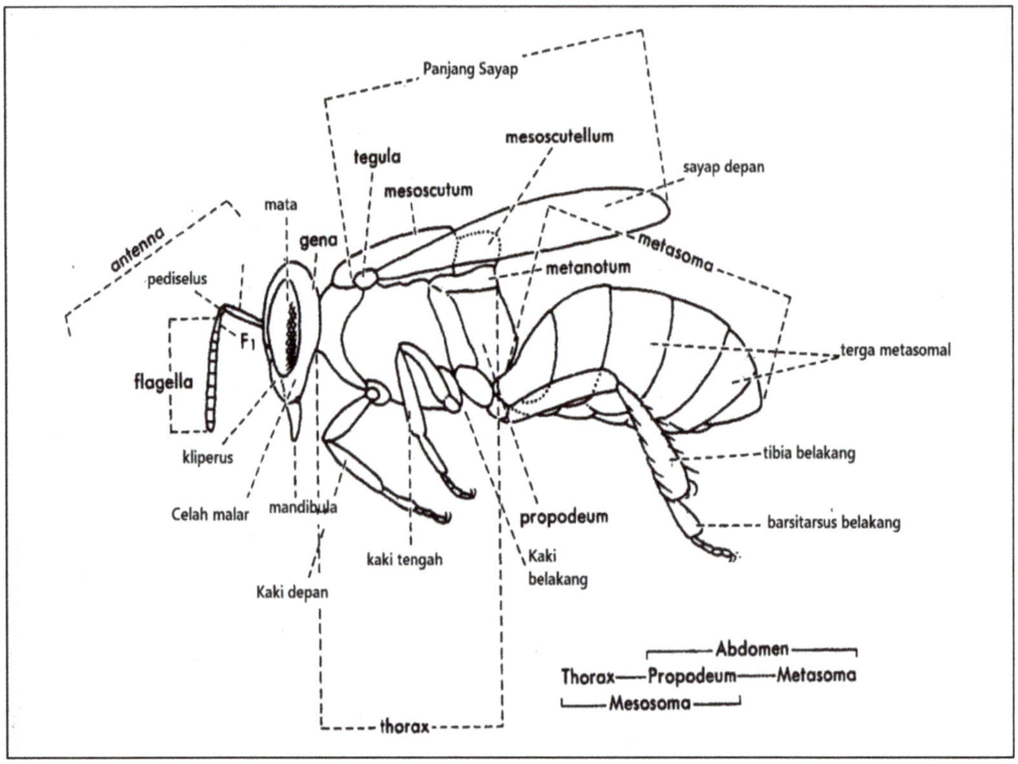

Egg-adult form of the stingless bee (*Tetragonula*)

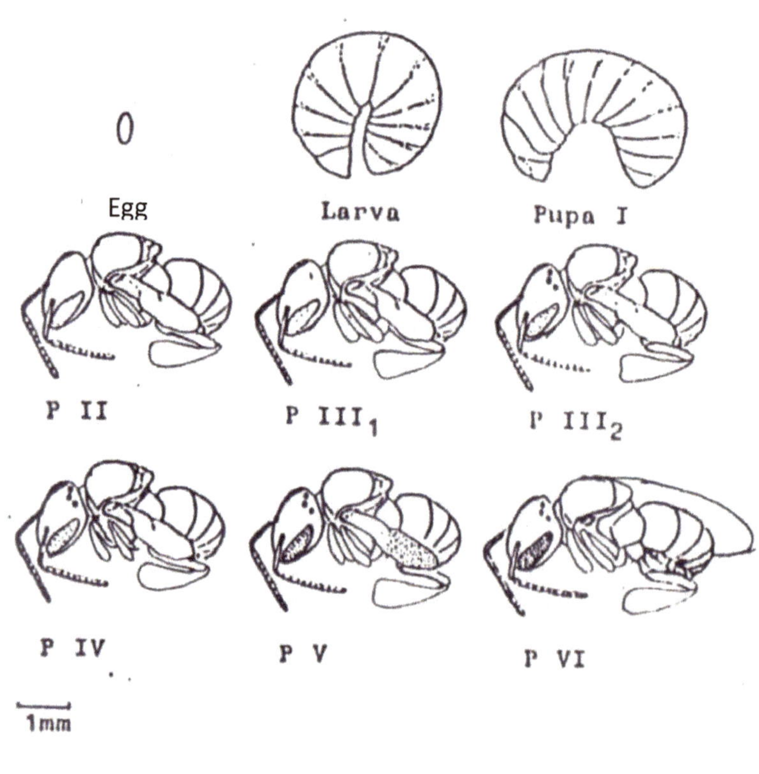

0.4 Incubation

The incubation period in workers of *Tetragonula minangkabau, Heterotrigona itama* and *Sundatrigona moorei,* compared with two Neotropical stingless bees and European honey bees

		Species								
Development Stage		*T. minangkabau*		*S. moreei*		*H. itama*		*Scaptotrigona postica **	*Melipona quadrifasciata ***	*Apis mellifera ****
		Start	Duration	Start	Duration	Start	Duration			
Egg		0	5.5	0	5.5	0	4.2	2	4.5 - 6.5	3
Larva	L1	5.5	3.0	5.5	1.0	4.2	5.0			
	L2	8.5	3.7	6.5	2.0	9.2	2.2			
	L3	12.2	2.8	8.5	7.0	11.4	3.2			
	Total		9.5		10.0		10.4	13	7 - 8	6 (5)
Pupa	P1	15.0	9.3	15.5	1.0	14.6	7.2			
	P2	24.3	4.3	16.5	5.3	21.8	2.7			
	P 3 1	28.6	3.2	21.8	3.7	24.5	4.4			
	P 3 2	31.8	2.4	25.5	12.0	28.9	7.5			
	P4	34.2	4.8	37.5	4.0	36.4	3.1			
	P5	39.0	2.6	41.5	3.0	39.5	5.0			
	P6	41.6	0.4	44.5	2.0	44.5	2.0			
	Total		27.0		31.0		31.9	31 - 33	20.5 - 23.5	12
Emergence		42.0		46.5		46.5		46 - 48	34 - 37	21 (20)

0.5 Pigmentation

Pigmentation period in workers of *T. minangkabau, H. itama* and *S. moorei.*

Development Stage		Species					
		T. minangkabau		*S. moreei*		*H. itama*	
		Start	Duration	Start	Duration	Start	Duration
Pigmentation changes in workers	Ia	0.0	1.3	0	0.75	0	0.3
	Ib	1.3	2.8	0.75	1.75	0.25	0.25
	II a	4.1	2.5	2.5	2.0	0.5	1.6
	II b	6.6	4.9	4.5	4.0	2.1	4.9
	III a	11.5	4.9	8.5	3.0	7.0	5.2
	III b	16.4	3.6	11.5	2.0	12.2	?
	IV a	20.0		13.5	4.0	?	?
	IV b			17.5		?	

Pigmentation Changes in Stingless Bee (*T. minangkabau*) Workers

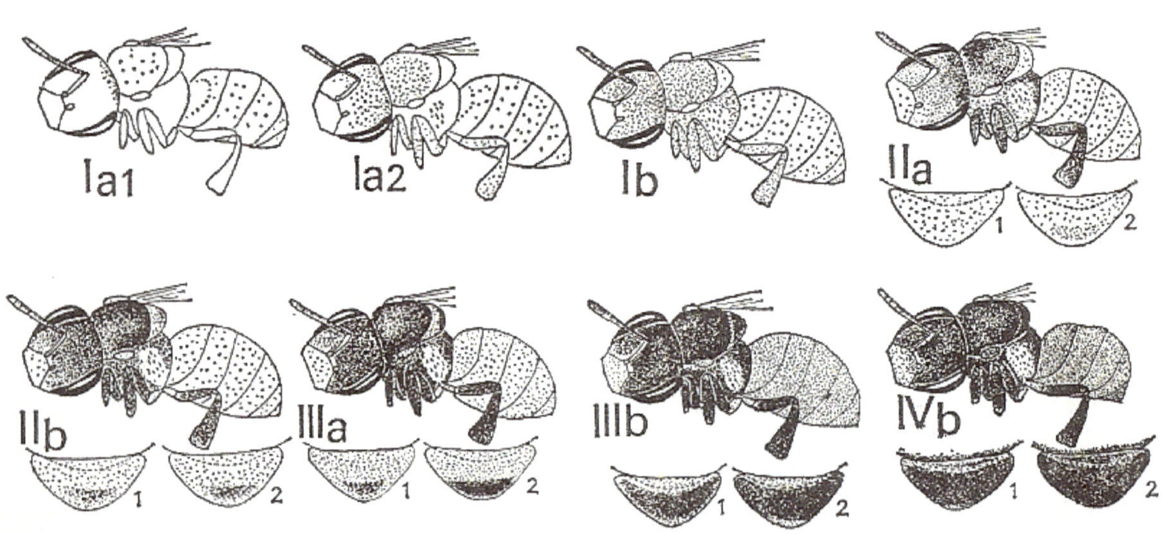

0.6 Nesting and structure

Stingless bee worker division of labour

- Core brood carers - generally take care of brood in cells and participate in the process of queen oviposition
- Peripheral and subsidiary brood carers/peripheral cell nurses, making food storage pots, nest entrance structures, and transporting and collecting garbage
- Foragers/ food seekers - do not participate in tasks in the nest and start looking for food/foraging earlier than other workers in the nest
- General workers - without a clear division of labour

Colony conditions can affect the division of the labour system

Stingless Bee Nesting Hosts

- Bamboo internode cavity
- Tree trunk hollows
- Foundation and wooden house wall cavities
- Cavity in Pillars
- Rock crevices
- Subterranean tree buttress roots
- Termite mounds
- Arboreal ant nests

***Heterotrigona itama* (below left) and *Wallacetrigona incisa* (below right) nest architecture**

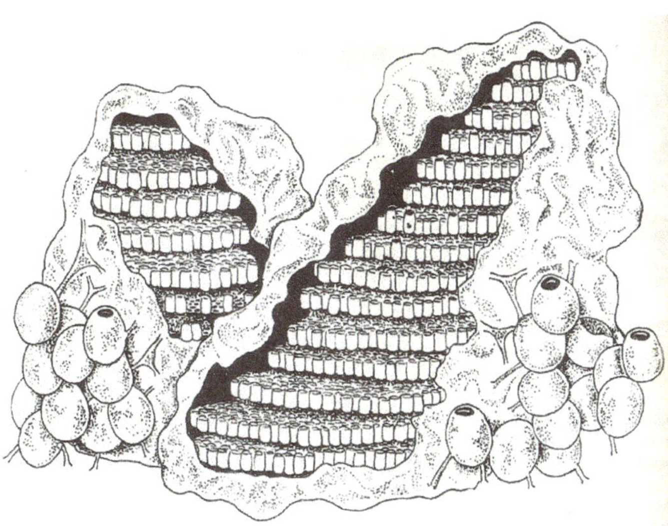

T. minangkabau (below left) and _S. moorei_ (below right) attending cells and their queen.

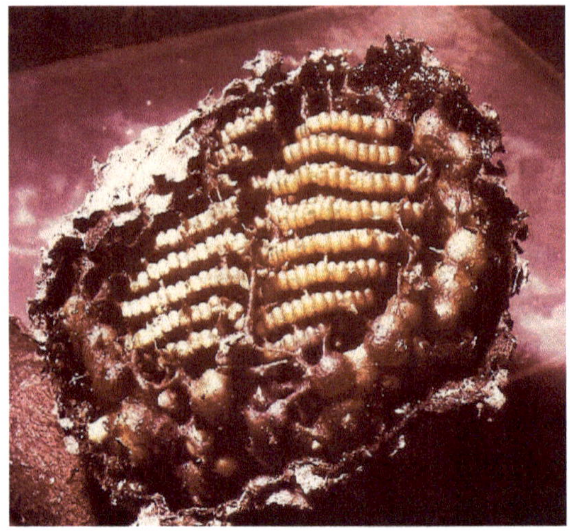

T. 26inangkabau Nest in Bamboo cavity

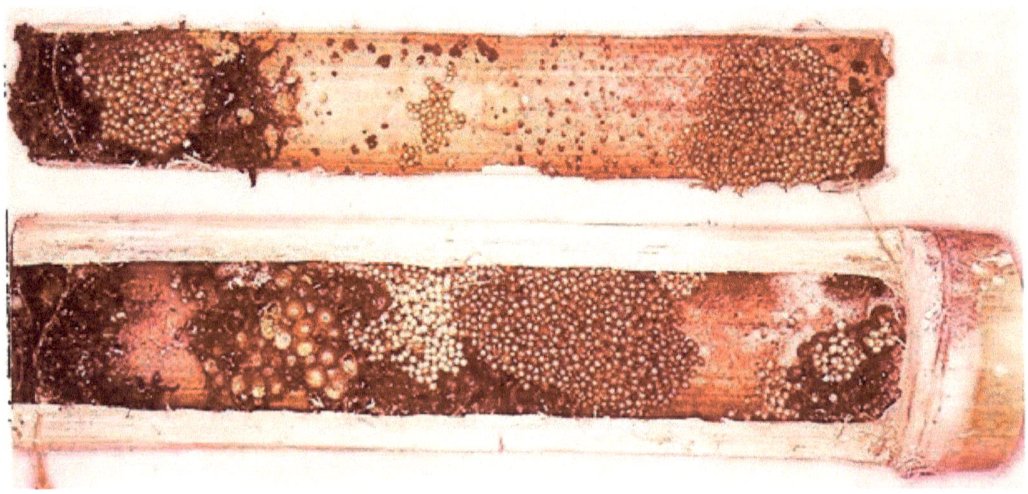

T. 26inangkabau Young Nest Architecture

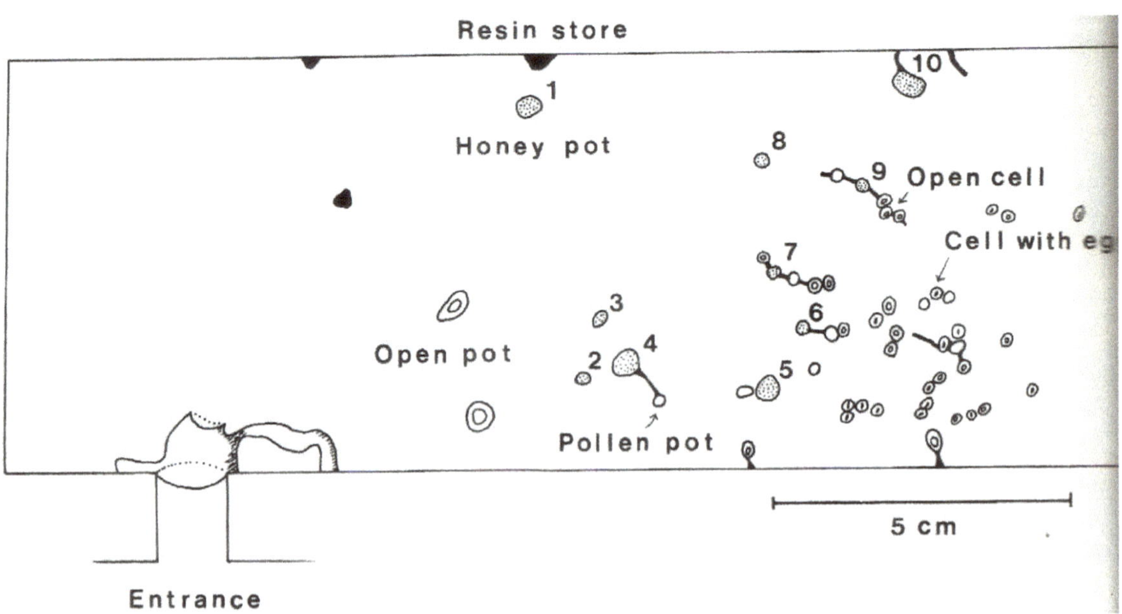

Illustration of *T. 27inangkabau* Nest in Bamboo internode

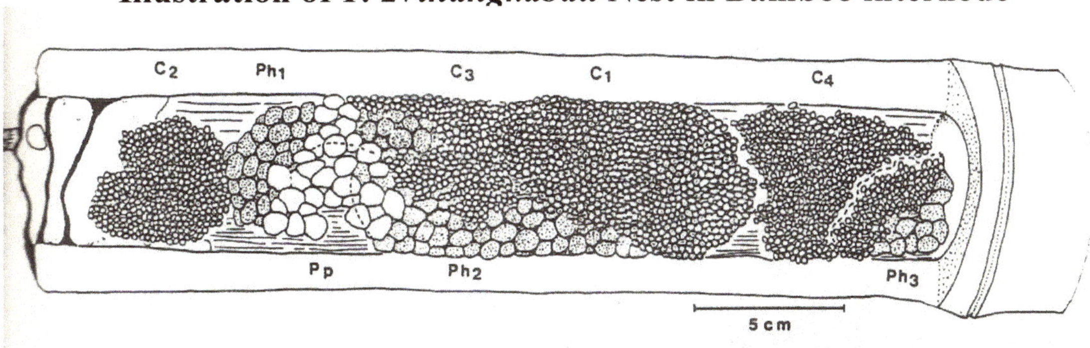

S. moorei colony in ants' arboreal nest

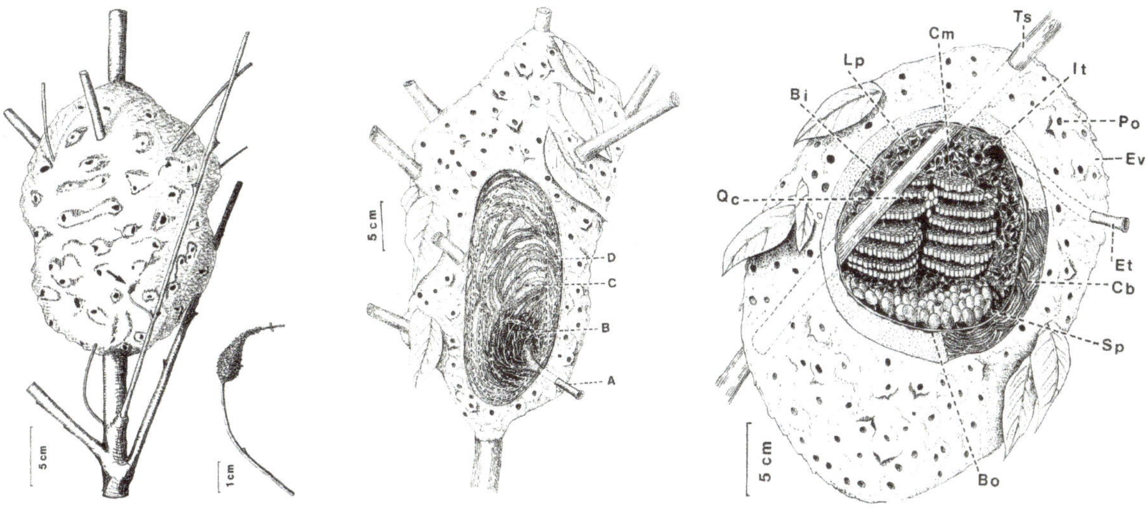

0.7 Meliponiculture

Stingless bee nest architecture with farming methods

8

Stingless Bee Functions/Benefits

- Pollination agent
- Honey Production (6.5 kg/yr; *Apis* 24 kg/yr)
- Pot Pollen
- Propolis (5.8 kg/yr; *Apis* > 1 kg/yr)

Utilization of Stingless Bees

<u>Pollination agent</u>

74% of flower-visiting insects *Meliponini* and *Apini* – Small size, generally pollinator in forests and small flowering plants.

<u>Honey Production</u>

A little honey, but the quality is the same as *Apis*. Sugar content 70-75%; Honey pots are different from pollen pots, so they are easy to harvest.

0.8 Products

Needs And Products Produced by Stingless Bees

No	Kebutuhan	Kegunaan	Produk
1	Pollen	Food for larva	Bee bread/Pot-pollen
2	Nectar	Energy source	Honey
3	Resin, gum, sap & latex (vegetative exudates)	Batumen (the inner layer of the nest cavity), involucrum (protective layer of brood clusters), cerumen (a mixture of propolis and wax to make nest pillars, honey and pollen pots), brood cells (usually made of cerumen), and nest entrances (mixed with sand/soil).	Propolis dan geopropolis
4	*Honeydew* (from *aphids, treehoppers* and *scale insects)*	Energy source	Honey
5	Minerals	Macronutrients, e.g., salts	-
6	Sand, soil, loam, mud, RESIN	Nest entrance	Geopropolis
7	Water	To cool down, *brood chamber*	-
8	Flower oil (collected from under the cuticle of flowers) there is no 'oil', it is gum or exudate for protection, like resin	Gives a distinctive aroma in the nest	Essential oils for Aromatherapy

- The vegetative exudates taken by the worker bees contain sap processed into black, yellow or dark brown propolis. The bees add salivary secretions and wax to make 'propolis'
- temperature 25 – 24 ^{0}C solid, soft, flexible and very sticky
- temperature less than 15 ^{0}C becomes solid but brittle
- temperatures above 45 ^{0}C become sticky and gummy
- temperature 60-70 ^{0}C becomes fluid
- after 100 ^{0}C will melt (fluid = melted)

Various types of plants influence the quality of propolis. Propolis cannot be consumed directly, and it must go through a process (inc. sterilization). It is already sterile. It kills bacteria and fungi. No process sterilizes propolis. Every single plant has different resin and sap chemistry.

0.9 Key to Worker Genera

Composite Algorithm Key to Indo-Malayan Meliponine Worker Genera

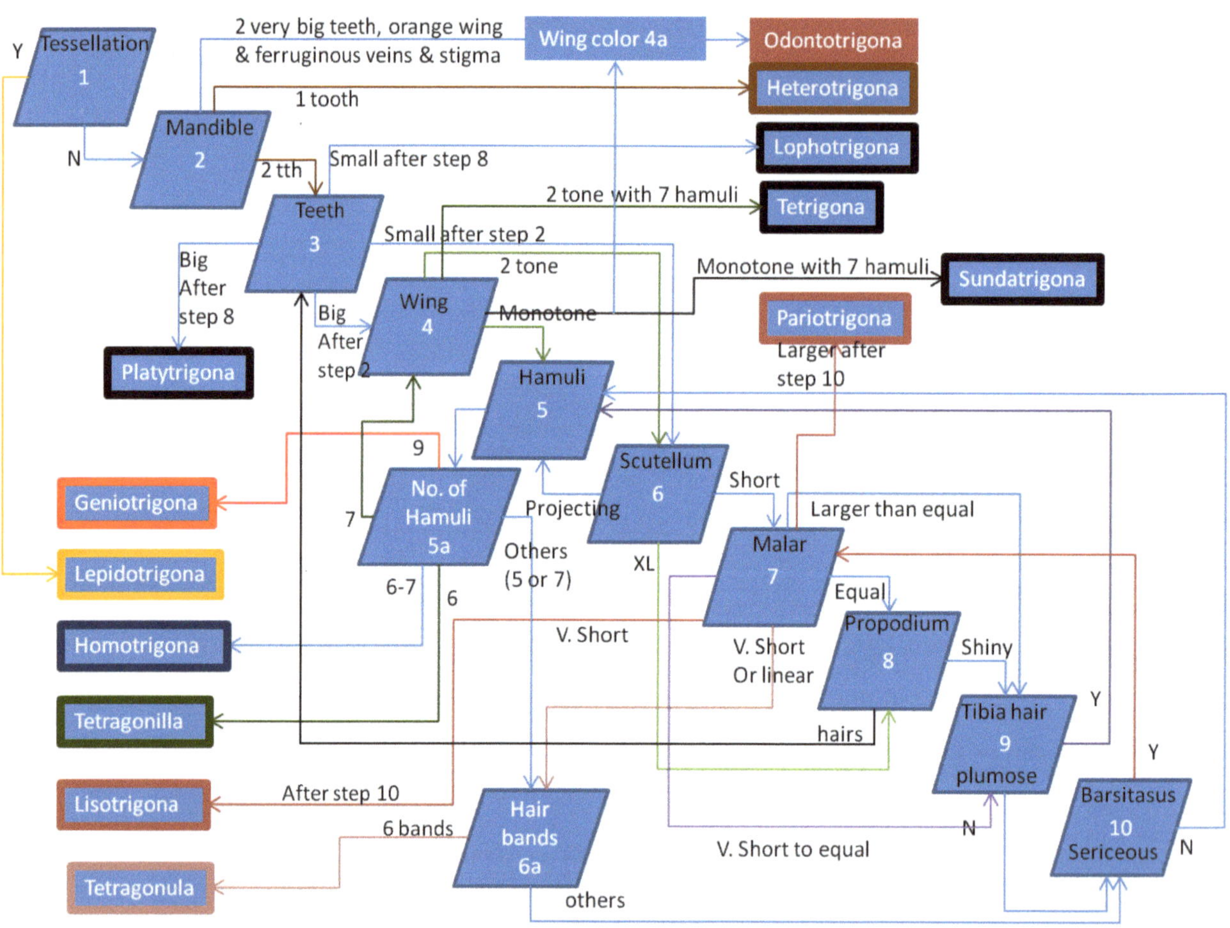

Source: Abu Hassan Jalil (Akademi Kelulut Malaysia)

0.10 References on Biodiversity of Indonesian Stingless Bees

Ascher, J., & Pickering. (28 Feb 2018). *Discover Life bee species guide and world checklist.* Retrieved from Sahulotrigona: http://www.discoverlife.org/mp/20p?see=Sahulotrigona&guide=1&flags=HAS:subgenus:

Ascher, J., & Pickering. (28 Feb 2018). *Discover Life bee species guide and world checklist.* Retrieved from Wallacetrigona: http://www.discoverlife.org/mp/20q?search=Wallacetrigona&guide=1&flags=HAS:subgenus

Ascher, J., & Pickering. (28 Feb 2018). *Discover Life bee species guide and world checklist.* Retrieved from Austroplebeia: http://www.discoverlife.org/mp/20q?search=Austroplebeia&guide=1&flags=HAS:subgenus:

Ascher, J., & Pickering. (29 Feb 2018). *Discover Life bee species guide and world checklist.* Retrieved from Tetragonula: http://www.discoverlife.org/mp/20q?search=Tetragonula&guide=1&flags=HAS:subgenus:

Ascher, J., & Pickering. (28 Feb 2018). *Discover Life bee species guide and world checklist.* Retrieved from Papuatrigona: http://www.discoverlife.org/mp/20q?search=Papuatrigona&flags=subgenus:

Baja, S. M., & Mostoles, M. D. (2017). COLONY CHARACTERIZATION AND AMPLICON SEQUENCING OF STINGLESS BEES (Tetragonula biroi). In A. H. Jalil, *Handbook of Meliponiculture vol 2.* Akademi Kelulut Malaysia.

Birt., M. (n.d.). The *Trigona* Species (Stingless Bees). Their roll in higher plant pollination. *12391833933.*

Chanchao, C. (2009). Antimicrobial Activity By T*rigona Laeviceps* (Stingless Bee) Honey From Thailand. *Pak J Med Sci April - June 2009 (Part-II) Vol. 25 No. 3*, 364-369.

Chandra, G. (n.d.). *Wallace's Line, Weber's Line and Wallacea.* Retrieved from IASZoology.com: http://www.iaszoology.com/wallacea/

Chinajariyawong, A., & Saiboon, S. (n.d.). Culture of the stingless bee, *Trigona laeviceps* Smith (Hymenoptera: Apidae). *Institute of Agricultural Technology, Walailak University.*

Chuttong, B., & Burgett, M. (2016). Biometric studies of the stingless bee *Tetragonula laeviceps* species complex (Apidae: Meliponini) in SE Asia (Thailand). *1st International Meliponine Conference.*

Ciar, R. R., & et al. (n.d.). Foraging Behavior Of Stingless Bees (*Tetragonula Biroi* Friese): Distance, Direction And Height Of Preferred Food Source. *UPLB BEE Program.*

Cimu, I. (1960). Results of bee pollination of Sunflowers. *Apicultura. 33(1) :* , 18-20.

Cortopassi-Laurino, M., & al, e. (2006). Global meliponiculture: challenges and opportunities. *Apidologie 37*, 275–292.

Darmawan, S. (n.d.). *Diversifikasi Produk Trigona Sp. Di Lombok.* Badan Litbang Kehutanan, Kementrian Kehutanan.

Dejtisakdi, W., & al., e. (n.d.). Melliferous plants for the stingless bees (Trigona collina Smith) in deciduous forest of Queen Sirikit Botanic Garden, Mae Rim, Chiang Mai Province. *Department Of Botany, Faculty Of Science, Kasetsart University.*

Dollin, A. E., Dollin, L. J., & Rasmussen, C. (2015). Australian and New Guinean Stingless Bees of the Genus Austroplebeia Moure (Hymenoptera: Apidae)—a revision. *Zootaxa.*

Dollin, A. E., Dollin, L. J., & Sakagami, S. F. (1997). Australian Stingless Bees of the Genus Trigona (Hymenoptera: Apidae). *CSIRO Australia Invertebrate Taxonomy.*

Eltz, T., & al., e. (2003). Nesting and Nest tress of stingless bees (Apidae; Meliponini) in lowland dipterocarp forest in Sabah, Malaysia with implications for forset management. *Forest Ecology and Management 172.*

Engel, M. S. (2000). *A Review of the Indo-Malayan Meliponine Genus Lisotrigona with Two New Species (Hymenoptera, Apidae).* American Museum Of Natural History.

Engel, M. S., Michener, C. D., & Boontop, Y. (2017). Notes On Southeast Asian Stingless Bees Of The Genus Tetragonula (Hymenoptera: Apidae), With The Description Of A New Species From Thailand. *American Museum Novitates.*

Inoue, T., Salmah, S., Abbas, I., & Yusuf, E. (1985). 1985. Foraging Behavior of Individual Workers and Foraging Dynamics of Colonies of Three Sumatran Stingless bees. . *Res. Popul. Ecol. 27:*, 373-392.

Jalil, A. (2016). *Meliponine Identifier Pictoruial Guide.* Akademi Kelulut Malaysia.

Jalil, A. H. (2019). *Cohabitation of Meliponines in Ants' or Termites' Arvoreal nests.* Aldemi Kelulut Malaysia Sdn Bhd.

Jalil, A. H., & Roubik(ed), D. W. (2019). *Filipino Meliponiculture and Beyond inc. MIMAROPA Tours.* Mindoro, The Philippines: Akademu Kelulut Malaysia.

Jalil, A., & Roubik(ed), D. (2016). *Handbook of Meliponiculture.* Akademi Kelulut Malaysia.

Jalil, A., & Roubik(ed), D. (2018). Handbook of Meliponiculture Vol 2. *Akademi Kelulut malaysia.*

Leonhardt, S. D., & al., e. (2010). Stingless Bees Use Terpenes as Olfactory Cues to Find Resin Sources. *Chem. Senses 35*, 603–611.

Michener, C. D. (1961). Observations on the nests and behaviour of *Trigona* in Australia and New Guinea (Hymenoptera: Apidae). . *American Museum Novitates 2026*, 1-46.

Michener, C. D. (2007). The Bees of the World 2nd ed. *The John Hopkins university press.*

Rasmussen, C. (2008). Catalogue of the Indo-Malayan/Australasian stingless bees (Hymenoptera: Apidae: Meliponini). *Zootaxa 1935.*

Rasmussen, C., & Camargo, J. M. (2008). A molecular phylogeny and the evolution of nest architecture and behavior in *Trigona* s.s. (Hymenoptera: Apidae: Meliponini). *Apidologie 39.*

Rasmussen, C., Thomas, J. C., & Engel, M. S. (2017). A New Genus of Eastern Hemisphere Stingless Bees (Hymenoptera: Apidae), with a Key to the Supraspecific Groups of Indomalayan and Australasian Meliponini. *American Museum Novitates.*

Riendriasari, S. D. (2014). Budidaya Dan Produk Perlebahan Trigona Spp Di Lombok, Nusa Tenggara Barat. *Prosiding Seminar Nasional* (pp. 213 -221). Jogja: University Club Universitas Gadjah Mada,.

Sakagami, S. F. (1975). Stingless Bees (Excl. *Tetragonula*) from the Continental Southeast Asia in the Collection of Bernice P. Bishop Museum, Honolulu (Hymenoptera, Apidae). *Jour. Fac. Sci. Hokkaido Univ., Ser. VI, Zool. 20 (1),.*

Sakagami, S. F. (1978). *Tetragonula* Stingless Bees of the Continental Asia and Sri Lanka (Hymenoptera, Apidae). *Jour. Fac. Sci. Hokkaido Univ. Ser. VZ, 2001. 21(2).*

Sakagami, S. F., & Inoue, T. (1985). Taxonomic Notes on Three Bicolorous *Tetragonula* Stingless Bees in Southeast Asia. *Kontyii, Tokyo, 53 (1): 174-189.*

Sakagami, S. F., Inoue, T., & Salmah, S. (1990). Stingless Bees of Central Sumatra. *Sapporo; Hokkaido University Press.*

Sakagami, S. F., Inoue, T., Yamane, S., & Salmah, S. (1983). Nest architecture and colony composition of the Sumatran stingless bee *Trigona (Tetragona) laeviceps. Kontyu, 51(1):,* 100-111.

Sakagami, S. F., Inoue, T., Yamane, S., & salmah, S. (1989). Nests of the Myrmecophilous Stingless Bee, Trigona moorei: How do Bees Initiate Their Nest Within an Arboreal Ant Nest? *Biotropica, Vol. 21, No. 3.*

Schwarz, H. F. (1937). Bornean Stingless Bees Of The Genus *Trigona. Bulletin American Museum of Natural History.*

Schwarz, H. F. (1939). The Indo Malayan Species of Trigona. *Bulletin of AMNH.*

Section 1 - Exploring Sumatra

By Abu Hassan Jalil (AKM) & Dian Angriasari (UNAND, Padang)

Figure 1 Al Qorni-qorni farm entrance

Contents in Section 1

Introduction to Section 1

A trip report of Lampung -West Sumatera – Riau - East Aceh. The objective of the excursion was to explore the meliponiculture practices and take account of the diversity of species being domesticated for Honey production on the island of Sumatra. The planning was done about two months before the event. The visit locations were obtained from beekeepers who were willing to accommodate our visits, and they were contacted via social media interest groups. The aim was to study the available vegetation suitable for meliponiculture of different species.

Map 1 Proposed excursion routes for Sumatra Island.

Accommodations were prebooked from online travel sites. The routes were organized from flight schedules available from numerous airline sites. (See Map).

The participants of the Meliponiculture area in Lampung this time were AHJ, Haron Abdullah, Michael Nyagu Sabang, Bladel Tan Kuan Yew (Malaysia), Prof. Siti Salmah and Dian Anggria Sari (Indonesia, West Sumatra). Meliponiculture visits in Sumatra should start on July 22, 2018. However, the visit begins the next day due to some inevitable matters. With the assistance of the Lampung area activity coordinator, Habibi Yaksyallah, Hendy Susanto, Yahya Sulaiman. All visit activities and accommodations were carried out smoothly without any hitches.

Chapter 1-1

1-1.0 South Sumatra

Map 2 Map of Lampung routes

1-1.0.1 – Al-Qorni qorni Farm @ Kampung Kecapi, Kalianda, South Lampung.

Heri Damora, as the owner of the Al-qorni qorni farm developed with family members, became the first place for a Meliponiculture Sumatra visit. The owners get their meliponine colonies from the forest, which generally comes from Sampir wood (Host tree), and some buy nests from other areas such as West Sumatra. Heri Damora is also a farmer who likes to be creative, as seen from how he makes decorations for nesting toppings and making nest toppings to observe the wrinkles making nests. Farmers also understand trapping bees, using '*apus'* bamboo, a large diameter type of bamboo to trap bees.

Up until the date of visit, Heri Damora had 80 colonies, from the types of *Geniotrigona thoracica* and *Heterotrigona itama,* which are maintained in areas planted with Cocoa (Koko), Cloves (Cengkeh), Dipterocarp Trees (Pohon damar), Areca Palms (Betel nut, Pinang), Jackfruit (Cempedak), *Averrhoa bilimbi* (Belimbing buluh), *Gnetum gnemon* (Melinjo), *Durio zebithinus* (Durian), *Manihot esculenta* (Cassava), *Mangifera indica* (Mango), *Archidendron jiringa* (Jering/Jengkol), Arengga or Sago Palm (Enau/Aren) and Coffee (Kopi). The diversity of plant sources of sap and shade for nests is found quite a lot in the maintenance area, but plants that provide sources of honey and pollen, which are always flowering, and vegetation of shrubs and ground cover are still inadequate, so they must be added and maintained, to ensure availability for colony needs.

Figure 2 Visit to Al Qorni Qorni in Kalianda District in Lampung Province.

Figure 2 - From left: Ansurasta Razak (Adik), Michael Sabang, Prof Siti Salmah Idrus, AHJ, Hendi, Heri Damora.

Location of maintenance of Al Qorni-qorni farm, Kalianda (See Map 2)

Hive box made with glass sides for observation.

Figure 3 Entry Pillar of oversized log hosting a nest and decorated with palm spadix sheaths.

Figure 5 1.2M tall thoracica log hive amidst other hives in Al Qorni-qorni farm

Updates December 2018 @ Al Qorni Qorni, Lampung.

Figure 4 Elaborate main entrance to the well Beescaped farm.

Figure 6 Some assorted meliponine hives internal structure and external features.

Figure 7 2m log hive and view of Beescaped farm.

Seems like Al Qorni-qorni farm prides itself with large and tall log hives, maintaining the original cavity size as much a complete whole as possible. This method is their way of conserving the original colony size.

Figure 8 Miniature scaled models of Meliponine colony hives

a miniature custom house that I made for a beehive-like this

Figure 9 A miniature truck model container hive; Bottom left: Geniotrigona thoracica nest entrance of a scaled model house.

Would you mind explaining a little about the construction, the type of wood and the objective to display the architectural culture of each model? The background of the architect and how long each model is, and how the results of its use in the cultivation of kelulut in it.

This model is a honey truck with honey-filled in the tub. If it's outside, I'll give a protective roof like this again, sir, for the type of wood, I use teak, sir.

Teak cannot be exposed outside to rain and hot sun. Teak is very good for indoor furniture. Advantage of beautiful grains teak. Also, please tell me a little about the treatment of the wood for termite resistance and the strength of joining, mortise, etc.

Yes, that's right, sir, if you put it outside, we will give you a protective roof again, sir, but God willing, it can withstand it. Apart from varnish, I also coat them with melamine so that the wood can be more resistant and can be more durable

Gallery of other scale models

Figure 10 Different scale models – a novelty in beehive carpentry. The sailing boat hive imparts a honey export aspiration.

My *Lansium domesticum* var. *duku* upright log wooden pack cabinet, the contents of the *Geniotrigona thoracica* colony

1. The topping box (Honey super)
2. Nest entrance to the log cavity hive.
3. 1st hollowed-out square shelves for honey stock and empty bottles
4. 2nd hollowed-out square shelves for packaging and labelling material
5. Closed-door storage for harvest suction apparatus, batteries and miscellaneous items.

1-1.0.2 – Agus Petir @ Brajasakti, Wayjepara, East Lampung

Agus Pujianto or Facebook Agus Petir is a farmer who has long maintained kelulut since 2004, utilizing and trading meliponine honey for traditional medical needs. Agus Pujianto maintains many *Geniotrigona thoracica, Heterotrigona itama,* and *Tetrigona apicalis* at different beekeeping sites and

Figure 12 Heterotrigona itama and a couple of log hives at the side of a residence.

Figure 11 Visiting team with the hosts @ Agus's Residence in Brajasakti district,

several Meliponini-type colonies found nesting in former ant nests (identified as *Sundatrigona moorei*). He also has his woodworking workshop to make topping boxes with honey extraction compartments that he maintains.

Figure 14 Left: Former ant nest, where the bee colony was found, which is thought to be Sundatrigona moorei; Middle: Colonies that have been moved into bamboo. Right: Internal structure of the moorei in Bamboo.

Figure 13 Giving directions to the cutting method and on how to remove the stingless bee nests

1-1.0.3 – Ryan Petrich @ Hargomulyo, Sekampung, East Lampung.

We arrived at Hargomulya rather late, and it was already nightfall. Enthusiasts and farmers greeted this third visit at Hargomulyo. Edo Saputra, a young man who is active in developing and maintaining Kelulut in his area (ID FB: Ryan Petrich), is happy to be creative in designing nests with various innovations in the maintenance and harvesting of honey, kelulut. Edo Saputra maintains various types of kelulut obtained from various regions in Sumatra. He also brought a colony of stingless bees endemic to Sulawesi via Borneo, namely *Tetragonula* cf. *biroi*. The nest hunter, one of Edo Saputra's associates, mentioned that most of the bees are found in Mahogany (not native?) and Coconut trees.

Figure 15 Left photo shows Prof Siti Salmah (in green) and Abu Hassan Together with meliponiculture farmers in Hargomulyo

Based on this, a discussion was held on how to understand the behaviour of kelulut in nature and apply it to the maintenance of the stingless bee colonies following the original environment and how to harvest good results, and try to open up insight into how to develop native species rather than bring in species from outside the area whose behaviour it is not yet known whether it will interfere with the sustainability of native species.

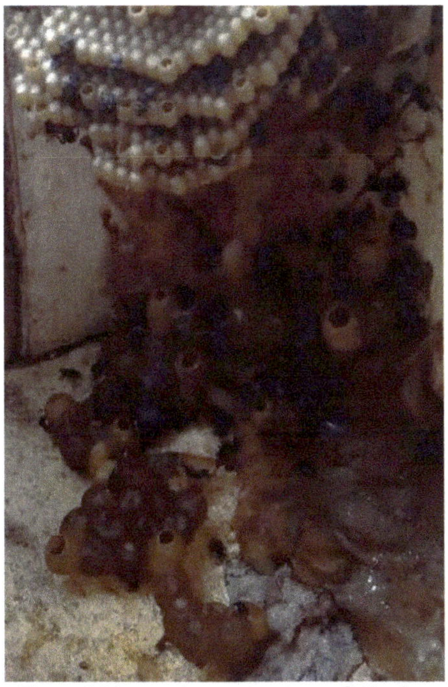

Figure 16 Internal nest structure of stingless bee within an arboreal Ants nest.

. The meliponiculture farmers in Ryan's group are Dimas, Sugi, Sam, Ridho, Cingceng, Gunawan, Sodri.

Based on visits from three locations in southern Lampung and East Lampung, it can be seen that the farmers have begun to understand how to raise good kelulut, but there is still much that needs to be improved from the management of kelulut maintenance that has been carried out. In addition, breeders only know utilizing the results in honey, while knowledge about processing and managing propolis is still very little known by farmers.

Here we found that some nests cohabited with arboreal ants' nests and posed some peculiar questions.

Questions posed to us from Ryan: What is the symbiotic role of ants with *Sundatrigona moorei*? Can they not be separated? Can we keep it in the wood box and succeed?

Answers: Yes, they can be separated. *Sundatrigona moorei* has been separated from ants successfully for about one year. It was done previously for research and publications (Prof. Siti).

Figure 17 Typical surroundings favoured by Sundatrigona moorei *at a higher altitude where it can be a little misty in the mornings and evenings.*

Habitat of *Sundatrigona moorei*, some mountains are a little foggy, cold temperatures, main pest *S. moorei* here are squirrels. Typically, the squirrel found is the 'plantain squirrel'. Living side by side with ferocious ants, squirrels eat the pollen, and in this *S. moorei* nest, I noticed a lot of pollen. When it is the harvesting season of the coffee, these bees become victims of farmers' insecticide because the ant nest is sprayed since it is felt the ants interfere with coffee harvest.

Figure 20 Typical arboreal ant nests cohabited by stingless bees, collected in the locality of a coffee plantation.

Figure 18 Typical arboreal ant nest cohabited by stingless bees Heterotrigona itama.

Figure 19 Typical arboreal ant nest cohabited by stingless bees Sundatrigona moorei

Figure 24 Sundatrigona moorei *coexisting with* Crematogaster *sp. ants in a coffee plantation*

Figure 23 Beebread: This amount of Beebread from 2 nests only

Figure 21 Typical layered brood cells of Sundatrigona moorei.

Figure 22 Plantain squirrel Callosciurus notatus (Citation: Duckworth, J. W. 2016) Downloaded on Jan 2nd 2019 from http://dx.doi.org/10.2305/IUCN.UK.2016-3RLTS.T3600A22254046.en.

8/3/2018 According to **Ryan Petrich**: We should encourage colony multiplication so that keepers are not dependent only on nature and minimal environmental damage. We tried doing eduction boxes attaching to wild colonies but mostly stolen by human thieves. There are still many forested areas, swarm trap boxes may be effective, if in our place the forest is gone, all of the farmland is already there if I don't think cutting in a protected forest is a problem, need to forge selective felling.

The case is the same as in Kalimantan. Hundreds of hectares of forest in question are made into oil palm land, which does not contribute greatly to the benefits of the surrounding communities.

Ryan Petrich: Hopefully, let's unite, learn, and exchange experiences. In the future, we can be independent, independent in breeding colonies, without us taking from nature again for propagation. I'm sure we will see someday; it must be changed.

[19/12/2018 'WhatsApp' forum on *S. moorei*]

Figure 25 Pak Sadiki's box hives of Sundatrigona moorei in Bengkulu, Sumatra (see Forum transcript below).

Sudaki Klanceng: From my observation, *S. moorei* tends to produce more queen eggs ... if the colony is strong, the number of queen egg cells are more than 15 and is broken down into 2/3 depending on the size of the colony where the fragments are originally located there is a queen ... wait until around one month usually the new queen appears in the colony without the queen (which has been given the queen's egg).

Gunawan Lampung: Although there are still five egg layers, there are four eggs from the candidate queen,

Sudaki Klanceng: In addition, prepare other colonies for egg donors if, at any time, the development is not according to plan. My experience is one colony which is broken up preparation two donor colonies, needs to be prepared with supporting vegetation, as I used to plant *Antigonon* flowers and sunflowers around the colonies other than existing plants.

Sudaki Klanceng: If you feel it is still lacking, I will feed meliponine honey / other bees with a little water mixed in.

AHJ: Can you get the percentage that succeeds?

Sudaki Klanceng: That was the first trial ... 70% triumphed.

Sudaki Klanceng: I don't understand beekeeping much, learning from numerous failures rather than results, only if the problem breaks more Alhamdulillah ever triumphed, it happened it might be a coincidence/

Gunawan Lampung: I only have one colony.

Sudaki Klanceng: Be patient, sir, until the colonies are really strong, and they will fail... as long as there are a lot of eggs, God willing, they can be saved even without donors. Donors are prepared if there are some; always better than if there are none. I want to try just with minimal, but usually, it's more successful if more.

Gunawan Lampung: Yes sir, I wait for this one colony to get strong, "where I have been looking for again, I have not found even though there were a lot of people here.

AHJ: From the beginning, how many colonies and still alive in the nests until now? Honest, sir?

Ryan Petrich: Let's be honest, in my experience, *Sundatrigona moorei* won't last long in an artificial wooden box. That's why they have never been found nesting in wooden cavities or trees. They have their reasons for living alongside ants.

AHJ: Thank you for the honest brief!

Ryan Petrich: Maybe Pak Abu (AHJ) can explain why *S. moorei* is always found inside the ants' nest, like as though they have a symbiotic existence between bees and ants, so nobody will keep on trying and try again, ending with big mistakes.

Agus Petir: But trying it is better, and understand the results?

AHJ: Ever tried and failed, it's better to give the correct info. Trying to deceive only gives false hope to other farmers. Every failure is a great loss to the environment and the treasures of our earth. Reason *S. moorei* made in the ant's nest (specifically *Crematogaster* sp.) is due to wall insulation found in arboreal ant nests (meaning nests in trees).

Agus Petir: Try and try until it works, and hopefully get better results.

Ryan Petrich: I think it is enough with my experience to damage dozens of *S. moorei* nests or learn from other people's experiences such as Pak Sudaki, but if you have strong reasons or new ideas that are not the same as conventional hives like you already have, can be considered.

AHJ: This is the safest way to do it, without damaging the walls of the ant ... can you just place the toping box above with the small hole above for them to go up ... if you don't go up (closed again, just let them) they will rise when they are available (as per photo of Tasby Amit's attempt in Brunei, see photo).

Agus Petir: I also don't want to damage or add, that's all, I have caused a lot of casualties.

AHJ: The walls are full of fine pores that can make the nest "breathe" ... it will not continue perfectly without proper ventilation.

Ryan Petrich: I have some information again about *S. moorei*. The symbiosis between ants and bees increasingly convinces me. I have tried cleaning ants by hanging them (not giving ants access to find food). Ants eventually die and run out. Not long after ants die, the bee's colony flies away, leaving the brood. How do you respond to this Mr. Abu Hasan?

AHJ: This is newly added info. The symbiosis was suspect (hypothetical) but never proven. Anyway, according to Prof Siti Salmah, during all her years of research on *moorei*, she has never seen *moorei* in nature in a trunk cavity or limestone rock cavity. So, there must be some attractant of the ants' nest for the bees to infiltrate and domicile in it!

Ryan Petrich: Nahh. If they could talk, I would ask them directly why they compact nest in the ant nest? Or, they may have protection from predators.

AHJ: Protection can also be possible, but ants and termites also nest in the ground. However, the *S. moorei* bees always follow aboveground ... they are never subterranean.

Prof Siti Salmah: In nature, *S. moorei* used only arboreal ant nest *Crematogaster* sp.

Figure 26 Tasby Amit's attempt to affix a topping box on an arboreal nest.

[19/12/2018 'Forum WhatsApp S. moorei]

Ryan Petrich on his current Meliponary status

1. the source of colonies from the surrounding area

2. harvesting is squeeze and sieve method (taking 60-70% only with 40-30% left behind)

3. harvest once per 2 months - 3 months (conditionally depending on the weather, if the weather is warm. harvest bimonthly; during the rainy season, it is not harvested)

4. honey harvest: plants around the dominant coconut, grasses (agrarian), *Asystasia gangetica*, cornfield, average yield per hive 40-50 ml

Species: *Heterotrigona itama*

1. Number of nests of *itama* gathered from the neighbourhood is about 40 logs (eastern Lampung)

2. Harvesting is by the suction system (vacuum pump)

3. Frequency of harvest is once per 10 days-2 weeks in the Kalianda environment and harvest frequency of 20 days - once a month in the coconut garden environment

4 Honey harvest: coconut, *Calliandra*, *Antigonon*, *Xanthostemon*, in the Kalianda neighbourhood, 500 ml-700 ml/month in the coconut environment average 400 -500 ml per month

Chapter 1-2

1-2.0 West & Central Sumatra

Map 3 Second Leg - Routes of road trips

1-2.0.1 – Padang, West Sumatra

We arrived in Padang, West Sumatra from Lampung (we had to transit in Jakarta). We parted company with Michael Nyagu Sabang, Bladel Tan Kuan Yew in Lampung as they had to return to Malaysia. In Padang, we met with other Malaysians Nur Raziana Ramli and Zaidi Bo and picked up Mohd Fadhli Sulaiman at the airport. Our arrival in Padang was primarily aimed at looking for *Tetragonula minangkabau*.

Figure 27 Arrival at the Minangkabau Airport, Padang, Sumbar.

We (the Malaysian group, Haron Abdullah, Mohd Fadhli Sulaiman, Nur Raziana Ramli & Zaidi Bo) together with (the Indonesian group, Prof Siti, Anggariani & Dr. Jasmi) continued to Pariaman to visit a bee farm there.

1-2.0.2 - Muchlisin @ Koto Dalam Barat, Padang Sago, Pariaman, West Sumatera.

Pak Muchlisin, a farmer in the city of Pariaman, wants to start and develop galo-galo (kelulut) farms in his village. Muchlisin maintains several galo-galo species in a simple bamboo hut, including *Heterotrigona itama* and several *Tetragonula* species. This activity was of great interest to Malaysians, namely Nur Raziana and Zaidi from the Malaysian Kelulut Academy. Here we interacted with the local community and discussed how they began the development of galo-galo farms in Koto Dalam.

Figure 29 Some of the internal structures left in bamboo, right in Box hive.

Figure 28 Together with Dr. Jasmi and Muchlisin, and some participants of Madu Kelompok Tani Beringin Jaya. Right photo showing the bamboo hives hung on stilts in Pak Muchlisin's Meliponary.

Their predicament is that the *Tetragonula* colonies of Muchlisin were not identified yet.

There are several species of kelulut and may be categorized into five sizes. The smallest is given the XS code, so on S, M, & L, and the biggest code is XL. Following the list of discoveries of *Tetragonula* in Sumatra, the five latest sizes are as follows:

(The measurement compared was HTL, i.e., the hind tibia length based on Sakagami 1975)

Tetragonula sp.	HTL
(XS) fuscobalteata	1.321
(S) minangkabau	1.556
(M) laeviceps	1.797
(L) drescheri	1.803
(XL) reepeni	2.336

Figure 30 Bamboo hives of Tetragonula *spp. affixed to stilts to accommodate within the compact shelter.*

Figure 31 Internal nest structure of nests in wooden boxes.

1-2.0.3 - Biology Department of Andalas University @ Limau Manih, Pauh, Padang, W. Sumatra.

The next visit was held at the Andalas Biology Department, with Mohammad Fahdli (Melaka) and Ariff Amri and Fadzilah from the Malaysian Kelulut Academy. Here AHJ gave a Public Lecture, and discussions about "Types and Potential of Stingless Bee in Southeast Asia" were held, attended by lecturers, students, researchers, breeders and practitioners in West Sumatra. After the public lecture was held, followed by a laboratory visit and a galo-galo nest survey in the Biology Research and Forest

Figure 32 Public Lecture and discussion at the Andalas Biology Department

Development, Andalas University.

Such meetings are important and need to be carried out continuously because academics, researchers and universities liaison farmers who need assistance and direction from the service and the local government in developing galo-galo farms and industrial products LIKE WHAT? produced by meliponines.

1-2.0.4 - Erfit @Tabek Simabur, Pariangan, Kabupaten Tanah Datar, Batusangkar, W. Sumatra.

Figure 33 Top left: Group photo of Malaysian visitors with Hosts (standing from left: Zaidi, Nur Raziana, Erfit's assistant, Pak Erfit, Ariff, Haron, Fadhli, seated from left: Prof Siti Salmah, Abu Hassan, Fadzilah & her 3 children); Bottom Left: Hollowed out Sampir wood log hive. Right: Sampir Tree (Oncosperma sp.)

The journey continues to Batusangkar, heading to the West Sumatra Galo-Galo House Group, where Erfit and friends of meliponiculturists and breeders from the area's Tabek Simabur are located. The location of the galo-galo maintenance is good enough, given the availability of food sources and sap, THEY DO NOT LIKE SAP- HONEYDEW? AND SAP IS NOT A GUM OR RESIN. and basic

Figure 34 Bamboo hives hanging from roof eaves.

vegetation shade and cover. In addition, the farmers' knowledge on the basic understanding of galo-galo life behaviour was quite sufficient. Erfit mentions that some of the colonies he owned came from other regions of Sumatra, some of which were found around his area, which were found nestled in Sampir wood (*Oncosperma* sp.).

Bamboo hives hanging at the roof eaves.

Figure 36 Tetragonilla atripes*, in Pariangan, Tanah Datar.*

Figure 36 Stacked log hives (Sampir wood)

1-2.0.5 - Frans Domo & Kelompok Madu Galo-Galo Kuok Lestari

In the late evening, the visiting team arrived at Kuok Emas River, West Bangkinang, Bangkinang, Riau, Kelulut Honey Group members – Galo galo Kuok Lestari, Safrans or better known as Frans Domo waited for the team at his residence. After a short break and being treated to typical Riau food, Bolu Kemojo and Kolak Ubi, the team immediately moved behind the house to discuss the management of the kelulut.

Figure 37 Observing Frans's honey harvesting

Figure 38 We were greeted with a garden picnic, where we proceeded with the introduction to the group members, Pak Adri was present and invited us for a discussion with the group.

At this location, it was seen that farmers still lacked information about how to manage and maintain kelulut well, as seen from the availability of inadequate food sources and kelulut and tending to barren maintenance locations.

He works with trap nests under Rubber Trees (*Hevea brasiliensis*). Here we can observe the extrafloral nectary of *Hevea brasiliensis*. Come harvest time, and the whole family gets involved, including his young child.

Figure 39 Top photos of internal structures of the toping boxes, Bottom left: bees foraging Rubber tree leaves' nectar. Bottom centre & right: Swarm trap containers made from 2,5 l plastic bottles.

1-2.0.6 - Ismail Arif @ Segati, Langgam, Kabupaten Pelalawan, Kampar, Riau.

Figure 40 Ismail Arif's Honey Label

Travelling from Kuok, guided directly by Ismail Arif and Adri Rahmadi, the next visit to Langgam, the Research Institute for the Development of the Stingless Bee, PT. RAPP-APRIL, where Ismail Arif and his colleagues maintain their colony of kelulut as a form of community service from PT. RAPP-APRIL. The team crossed the Kampar River to reach the maintenance site, using Pompong or Pontoon in Rantau Kasih's

Figure 42 Vehicle crossing using Pompong / Pontun in Rantau Kasih on the way to the meliponiculture site.

Figure 41 Left photo: Sumatra Excursion Visit Team with Ismail Arif. Right photo: The spirit of these two elders makes us more confident to continue to teach how to cultivate bees to the community.

village. Pompong or Pontoon is a simple iron raft tied with ropes and pulleys and moving river currents.

After crossing, we must pass through Eucalyptus and Palm plantations before arriving at the location of the kelulut farm. Although the maintenance location seems arid, the products produced from the colonies of *H. itama* and *G. thoracica* owned by Ismail Arif are quite large, in 40 days 2L per topping, with enough food sources and sap[1], including Jackfruit, Jengkol, Coconut and Palm trees. Naturally, many colonies in nature are found in Jengkol tree trunks.

Figure 43 Few types of meliponine found at Pak Ismail's Beefarm.

Image of making sample hives for *G. thoracica*. When the colony was strong, we designed this hive to harvest honey and the manual multiplication process of colonies. All coupled with assurances that we can spread immediately to the community.

Figure 44 Left) Location of Maintenance of the Meliponine Colonies of Ismail Arif in Segati, Langgam; Right) Ismail Arif is a lecturer in the field of timber engineering business.

---

[1] Editor's note: Sap is a carbohydrate, much like gum. Resin is terpenoid. sap is a product of photosynthesis and it is sugary, phloem sap is essentially the same as honeydew, provided by Hemiptera to bees or ants.

1-2.0.7 - Hj Ridwan & Pak Joyco @ Desa Benteng Hilir, Mempura, Siak, Riau.

From the Langgam to Siak, the team moved towards the place of Joko Paryanto, or Joyco, Muhammad Ridwan, Dadang Sofyanto in the Pemuda Tani Taruna Jaya Group maintaining kelulut. From environmental conditions, the availability of food sources and resin, shade and basic vegetation cover has been fulfilled for the needs of the colonies in general. However, the number of colonies that are too many and the density of placement was too tight and is feared to reduce the productivity of the stingless bee colonies. Location of maintenance of meliponine colonies in Siak

Figure 46 "One of the objects that motivates us to plant flowers is the number of bees that approach the flower. Especially if you see the funnel like this." said Ismail.

Figure 45 Visiting Muhammad Ridwan (in blue suit & white cap) in his residence garden & Dadang Sofyanto in the Pemuda Tani Taruna Jaya Group.

The Visit Team with the Kelulut Farmers from Langgam and Siak met at Mr. Muhammad Ridwan's house. Some of our team members visited the neighbour Mr. Joko Paryanto's, or Joyco next door.

Here we can see Joko Paryanto harvesting honey in his field located in the front garden of his residence.

Figure 47 At Pak Joko's place we examined a few colonies to observe the consequence of the difference in the available vegetation as compared to the neighbour's (Hj Ridwan) colonies.

1-2.0.8 - Rasyid & Tengku Fadhli @ Siak Sri Indra Pura.

Arrived at Siak Sri Indra Pura and visited Rasyid and his neighbour Tengku Mohd Fadhli. Both were just beginners in Meliponiculture, and between them, they had about 50 odd hives.

Figure 48 Visiting the neighbour Joko Paryanto aka Pak Joyco. Here we can see Joko Paryanto is harvesting honey in his field which is located in the front garden of his residence.

Figure 50 Top left: Rasyid briefing Prof Siti & Anggriani (striped blouse) Bottom left: group photo

Figure 49 Hives in his Banana farm

Figure 51 Some of the Log colonies. Top from left: leban (Vitex sp.), durian (Durio sp.), mahang (Macarana sp). Bottom from left: durian (Durio sp), village rambutan (Nephelium sp.),

1-2.0.9 Taman Hutan Raya UPT KPHP Minas

The location of the last visit in Riau took place at the Sultan Syarif Hasyim Forest Park (TAHURA), UPT KPHP Minas. With assistance from Pak Adri Rahmadi, the team met with the Field Coordinator TAHURA Minas, Pak Tumpal Simamora. The team discussed meliponiculture with him at the maintenance site. The number of colonies maintained in this Tahura is quite large, but the food sources always available are inadequate because of the vegetation dominated by tall trees with once-a-year flowering time and tight shade. The threat of monkey pests that disturb nests has also been observed and is difficult to overcome

Figure 52 Sultan Syarif Hasyim Forest Park (Tahura - Taman Hutan Raya), UPT KPHP, Minas.

because monkeys are protected animals in the region. Sun bears *(Helarctos malayanus)* are also another major problem. In addition, the installation of logs and toppings used is not following good maintenance rules, such as being placed too high or the size of a box that is too large or too small, not adjusted to the existing species.

Figure 54 Location of Meliponiculture in Tahura Minas Forest Reserve.

Figure 55 Video recording of mother and cub Bears who were discovered by Bpk Simamora (Field Coordinator) Minas Tahura KPHL (Riau) Monday August 13, 2018 at 6:00 p.m.

Figure 56 Lepidotrigona terminata, one of the species found in Tahura Minas Forest Reserve.

Figure 53 Visiting team - from left Tim pelawat – Haron Abdullah, Tumpal Simamora, Adri Rahmadi, AHJ & Dian Anggriani.

Aspirations for the place developed further into a tourist destination and a Holiday Camping ground. Renovations are underway, and elaborate planning has gone into it. Pak Adri also showed us his Lab and his Dehydration unit for Honey with high moisture content.

Chapter 1-3

1-3.0 - Aceh District, North Sumatra

Map 4 Aceh District, N. Sumatra

1-3.0.1 - Bakhtiar @ Blang Simpoh, East Aceh. On 29th July Morning, we departed from Binjai to Langsa and met up with "linot" (Acehnese term for meliponines) beekeepers of East Aceh. The beekeepers were Bakhtiar, Alrahman & Omjol before leaving for Blang Simpoh, where Bakhtiar's farm is. We (AHJ & Haron Abdullah) arrived in Medan on the 28th evening and was scheduled to meet with Pak Rahmat Hidayat, but was 'stood up', so we just put up the night in Binjai and made arrangements to hire some transportation for the next day. The attached write up is a translated blog written by some local East Aceh reporters.

Figure 57 Bahtiar with a Termite nest inhabited by Heterotrigona itama *in Blang Simpoh, East Aceh.*

Figure 58 AHJ Interacting with Bahtiar as Omjol & Rayhan looked on.

East Aceh - The Academic Research Team from Malaysia came to Aceh to share knowledge in honey cultivation, or in traditional languages referred to as 'linoet', which Bakhtiar is currently cultivating in Tualang Pateng village, East Perlak District, East Aceh.

On this occasion, Abu Hasan Abdul Jalil, a Malaysian stingless bee researcher accompanied by Raihan, a researcher from Syiah Kuala University, helped motivate the development of Linoet cultivation in East Aceh.

As previously known, Bahtiar has been running Linot cultivation since 2015 with a profit of up to millions of rupiah. Only with nests in logs and as many as 40 toping boxes, and with considerable ability, he continues to run the honey business.

Abu Hasan Abdul Jalil, the Malaysian Academy of Meliponiculture, in his written statement received

Serambi Timur

Serambi INDONESIA

Selasa 31 Juli 2018 | 11

Peneliti Malaysia Tinjau Peternakan Lebah Kelulut

IDI - Tim Peneliti dari akademisi Kelulut Malaysia, yang diketuai Abu Hassan Jalil, Minggu (27/7) meninjau lokasi peternakan lebah kelulut di Gampong Tualang Pateng, Kecamatan Peureulak Timur, Aceh Timur. Peternakan lebah kelulut tersebut merupakan usaha yang dirintis Bahtiar, warga setempat.

Lebah Kelulut atau dalam bahasa Aceh disebut *linot*, merupakan jenis lebah yang tidak bersengat yang juga menghasilkan madu sama seperti lebah madu.

Dalam kunjungannya, Abu Hassan Jalil yang ditemani Raihan, peneliti dari Universitas Syiah Kuala (Unsyiah), dan Charles selaku investor asal Sumatera Utara, berbagi pengetahuan tentang cara dan teknik mengembangkan peternakan lebah kelulut di Aceh Timur.

Cara beternak lebah kelulut yang dilakukan Bahtiar selama ini, menurut Abu Hassan, masih banyak yang perlu diperbaiki. Seperti pembuatan sarang yang baik dan benar, cara agar terhindar dari hama, dan parasit yang dapat mengganggu perkembangan kelulut.

"Budidaya lebah kelulut ini berpeluang dikembangkan menjadi usaha industri. Karena itu untuk mewujudkannya, perlu adanya bantuan dan dukungan dari Pemkab Aceh Timur melalui dinas terkait," jelas Abu Hassan.

Sementara, Bahtiar mengaku, beternak kelulut ini telah ia tekuni sejak 2015. Ia kini memiliki sekitar 40 sarang kelulut yang dipelihara di kebun sawitnya. Dari usaha itu, ia kini telah meraup keuntungan hingga jutaan rupiah.

"Madu dipanen selama tiga bulan sekali dengan harga jualnya Rp 500 ribu per kilogram," jelasnya.

Karena keterbatasan modal, Bahtiar berharap pemerintah atau investor bersedia mendukung usaha yang dikembangkannya tersebut. "Kami masih jauh ketinggalan dari peternak kelulut di Malaysia. Karena itu, kami harap dengan bantuan dan dukungan dari pemerintah, kami bisa diajarkan teknik beternak yang benar, serta terbukanya peluang penjualan yang menguntungkan peternak," harap Bahtiar. **(c49)**

TIM Peneliti dari Akademisi Kelulut Malaysia, yang diketuai oleh Abu Hassan Jalil, Minggu (27/7) meninjau lokasi peternakan lebah kelulut di Gampong Tualang Pateng, Kecamatan Peureulak Timur, Aceh Timur. Peternakan lebah kelulut tersebut selama ini dikelola oleh Bahtiar wa[rga] setempat.

BKPRMI Gela[r] Perkampungan R[...]

Figure 59 News clipping of Serambi Indonesia 31 July 2018.

by HARIAN ACEH.co.id on Sunday (07/29/2018), said that he went to Aceh to share knowledge and stay in touch so that Linoet meliponine breeders in East Aceh would be more developed.

Figure 60 The village community came out to join in the eventful visit to Bakhtiar's Meliponary.

"I came to East Aceh, wanted to share, share knowledge or exchange opinions and stay in touch so that honey cultivation of kelulut or 'linoet' as in the traditional language could develop in Aceh, especially in eastern Aceh," he said.

Furthermore, he added that kelulut cultivation has the potential to create jobs. So, this needs the attention of the regional government to become an organized business.

"I think if this cultivation cannot be developed into this industry, it is very disadvantageous. And if the cultivation of the linot is to be made into a kind of cottage industry, it needs attention from the relevant agencies regarding the cultivation of linoet, so that stingless bee culturing can develop.", Abu Hasan added.

Several things should be of concern to entrepreneurs in carrying out this business, especially the care of developing good bee housing, then handling pests that threaten the linot.

"This effort needs to be considered as well as making better Linoet nests, care for the place used when cultivating Linot must be maintained properly, so that it can produce good quality honey and eventually it can also boost the income of the community", said Abu Hasan Abdul Jalil, eager and optimistic.
Similarly, Bakhtiar believes that the Linoet cultivation activity he is currently developing has become a small business for honey cultivation by some people in East Aceh. He said he was very happy because the residents in East Aceh began to pay attention to this Linoet cultivation.
"Alhamdulillah, the results accumulatively, from every hive box we get up to 1 kg each of Kelulut honey (on average), and now the market is Rp.500,000 per kilo which is harvested every three months", Bahtiar said.
However, he said that some considerations for maintaining a business, such as development to increase income, ideally include the Aceh government's funding and investors' capital.
Meanwhile, AHJ, as a Researcher from the Malaysian Kelulut Academy, said that the cultivation of

Figure 61 Bakhtiar's Meliponary setup with some of the Hives and the location of Bakhtiar's farm adjacent to an oil palm estate in Blang Simpoh (see map).

kelulut carried out by Bahtiar needs development and making better hives to get more honey.
"There is still much work from cultivating kelulut in East Aceh, such as building hives that can last for at least ten years. This way, their businesses can be improved, and the results obtained can also be increased," he said in Pateng, East Aceh. Reports in Serambi Timur 31st July 2018 (see News clip above)

1-3.0.2 - Alrahman P. Pasi @ Julok, East Aceh.

We spent the night in Kota Idi (about midway between Peureulak and Julok) before visiting

Figure 63 Surprisingly found a luxury hotel in Kota Idi (closest to Julok east Aceh) Even had meeting facilities and fine dining.

Alrahman. We gathered to discuss our plans for the next day at The Royal Idi Hotel. The next day, we arrived in Julok and were brought to the Meliponiculture group's location.

Figure 62 Some of Alrahman's stingless bee colonies.

Figure 66 Vegetation and foraging activies of their stingless bees (mostly Heterotrigona itama*).*

Figure 64 Rest hut where formal and informal discussions are held. A typical traditional platform in every sense.

Figure 65 Meliponiculture at Alrahman at his residence.

Figure 68 Top left: The Meliponiculture group of Julok, East Aceh; Example of the groups' carpentry works.

Figure 67 G. Nest of thoracica *being invaded by a toad; Right photo: unidentified meliponine with clear contamination with possibly nearby wet paint.*

Figure 69 Alrahman with an extracted Geniotrigona thoracica *brood for reviving established hives; the* G. thoracica *nest entrance and the original log hive.*

East Aceh Alrahman Pasi Kind of species: *H. itama, G. thoracica*; ~ 50 nos. Topping size 30cm x 30cm x 6cm x 2cm. Harvest by suction machine per 30-40 days minimum yield of 500 ml maximum 1.8 litres. Rp 6-8m; BSF = 1 in attack; *Tetragonula* = 2 the yard of the residents rented areca nuts, guava, rambutan, rubber, palm oil, coconut, mango, acacia, *senduduk*, tagetes, and many shrubs and grasses. Result: 6-8 million per month not yet fixed. Eight million earnings for about 50 colonies. Those who are ready to harvest only 30 colonies are not yet optimal.

Figure 70 Intsia bijuga *Tree (Ipil plant)*

Meliponiculture in Aceh, North Sumatra, by Alrahman P. Pasi

Part of the uniqueness of Kelulut/Linoet honey is that its colour will be different in each colony, with each place and vegetation.

Figure 71 Instances of green honey harvested.

Bees are one of the living creatures classified as beneficial animals because bees do many things for humans, starting from producing honey, which is very good for human health.

Figure 72 Treatment and fabrication of "topping" boxes

Figure 73 Cosy setting for the bees and beekeepers.

Preparation to enter the harvest season while harvesting honey, we put on new topping boxes. In addition to healthy colonies and abundant vegetation, cleaning or changing toppings each period is one way to keep honey production stable and good in the harvest.

However, have you ever wondered what would happen if bees became extinct or no longer existed in the world? This concern is certainly possible. If bees are extinct, then there will be something that threatens humans.

1-3.0.3 Omjol @ Peureulak, East Aceh.

We were shown Omjol's collection of *Geniotrigona thoracica* hives on his farm upon arrival.

Figure 74 Geniotrigona thoracica *hives in Omjol's collection*

Figure 75 Scaled model miniature traditional house as stingless bee colony housing.

Scaled model miniature traditional house as stingless bee colony housing.

1-3.0.4 Rayhan Zaky @ Lokhseumawe, North Aceh

We later travelled to the borders of North Aceh, Lokhseumawe heading to Rayhan Zaky's house.

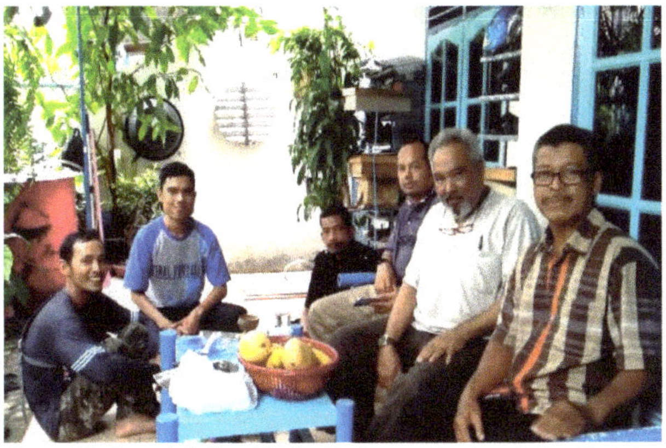

Figure 76 From left (Dari kiri): Alrahman, Rayhan Zaky, Omjol, Haron Abdullah, AHJ & Rayhan's Father.

Figure 77 Commercialised packaging or Ar-Rayhan Honey.

What is Ar-Rayhan Honey?

Ar-Rayhan Honey is a pure Linöt (Linot, linoet????) Honey is harvested without heating, pasteurization, or processing, to maintain its natural goodness. This honey is produced by bees from the Bee tribe Meliponini, namely bees with no sting and smaller body size than *Apis* (stinging) or even flies. These stingless bees have several names such as Linöt (Aceh), Kelulut (Malaysia, Kalimantan), Klanceng (Java), Galo-galo (Padang), Stingless Bee, Trigona, and others. Linöt bees are often overlooked by us, even though the efficacy of honey (i.e. linöt) produced is more efficacious than ordinary honey. Its small size allows linöt bees to access even a very small number of flowers, things that cannot be done by the other honey bees (*Apis* spp. / Stinging honey bees) so that the content in honey is more varied.

Figure 78 Rayhan Zaky is a graduate student who produces AR-RAYHAN Honey.

Figure 79 Arrangement of meliponine box hives at Rayhan's residence.

In the nest, linöt bees store their honey in pots as big as marbles made of propolis, in contrast to the stinging bees storing honey in the beeswax bag. Honey stored in a propolis bag will absorb antioxidant ingredients from propolis and make it honey at the level of "Medicinal Grade Honey" or a special honey treatment. This honey is commonly used by locals inland as a basic ingredient of traditional medicine and is often dubbed "The Mother of Medicines". (Rayhan Zaky, 2018)

MADU AR-RAYHAN

Section 1 Summary

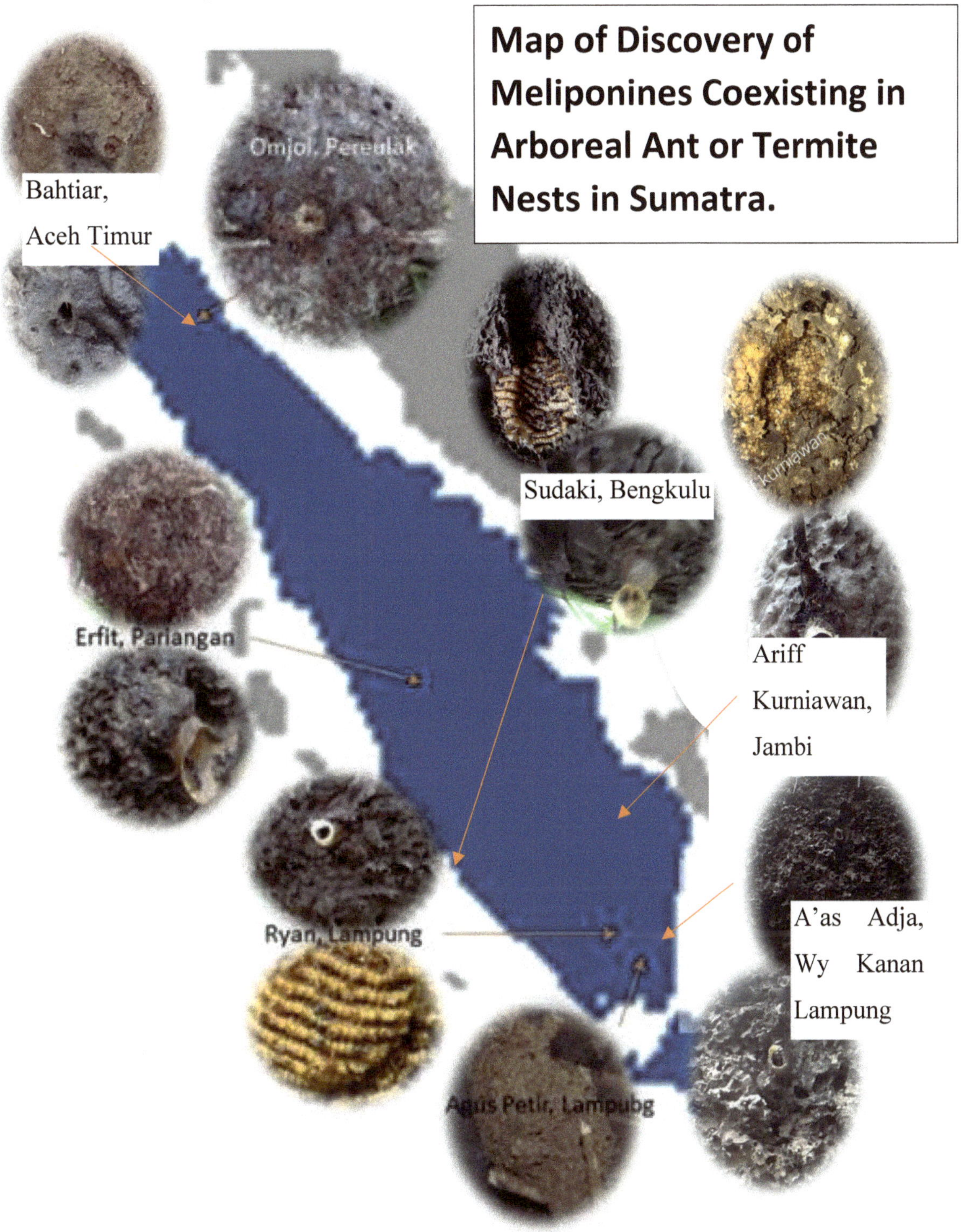

Map 5 Map of Discovery of Meliponines Coexisting in Arboreal Ants or Termites Nest in Sumatra.

Figure 80 Variations of Heterotrigona itama found in different regions of Sumatra, Indonesia

Figure 81 Variations of wing tip colour tones of Geniotrigona thoracica.

Variations of shapes and colors of entrance tubes of *Geniotrigona thoracica* in Sumatra

Figure 82 Variations of shapes and colours of entrance tubes of Geniotrigona thoracica in Sumatra. Part 1.

Figure 83 Variations of shapes and colours of entrance tubes of Geniotrigona thoracica *in Sumatra. Part 2.*

1-3.0.5 References for Section 1

Appanah, S., & Turnbull (eds), J. M. (1998). *A Review of Dipterocarps: Taxonomy, ecology and silviculture.* FRIM.

Ascher, J., & Pickering. (2018, Feb 29). *Discover Life bee species guide and world checklist.* Retrieved fromTetragonula: http://www.discoverlife.org/mp/20q?search=Tetragonula&guide=1&flags=HAS:subgenus:

Bingham, C. (1903). Diagnoses Of Aculeate Hymenoptera. *Fasciculi Malayenses.*

Boongird, M. &. (2004). A New Species of *Trigona* from Peninsular Thailand (Hymenoptera: Apidae: Meliponini).

Ciar, R. R., & et al. (n.d.). Foraging Behavior Of Stingless Bees (*Tetragonula biroi* Friese): Distance, Direction And Height Of Preferred Food Source. *UPLB BEE Program.*

Cockerell, T. D. (1919). The Social Bees of The Philippines Islands. *University of Colorado.*

Dollin, A. E., Dollin, L. J., & Rasmussen, C. (2015). Australian and New Guinean Stingless Bees of the Genus *Austroplebeia* Moure (Hymenoptera: Apidae)—a revision. *Zootaxa.*

Dollin, A. E., Dollin, L. J., & Sakagami, S. F. (1997). Australian Stingless Bees of the Genus *Trigona* (Hymenoptera: Apidae). *CSIRO Australia Invertebrate Taxonomy.*

Engel, M. S., & Rasmussen, C. (2017). A new subgenus of *Heterotrigona* from New Guinea (Hymenoptera: Apidae). *Journal of Melittology.*

Engel, M. S., Michener, C. D., & Boontop, Y. (2017). Notes on Southeast Asian Stingless Bees of the Genus Tetragonula (Hymenoptera: Apidae), with the Description of a New Species from Thailand. *American Museum Novitates.*

Heard, T. (1999). The Role Of Stingless Bees In Crop Pollination. *Annu. Rev. Entomol. 44:*, 183–206.

Henderson, A. (1986). *A Review of Pollination Studies in the Palmae.* The Botanical Review.

Jalil, A. (2014). *Beescape for Meliponines.* Singapore: Partridge Publishing.

Jalil, A. (2016). *Meliponine Identifier Pictorial Guide.* Akademi Kelulut Malaysia.

Jalil, A. (2016). *World Meliponine Etymology of Taxanomic Nomenclature.* Akademi Kelulut Malaysia.

Jalil, A., & Roubik D. W. (ed). (2016). *Handbook of Meliponiculture.* Akademi Kelulut Malaysia.

Lamerkabel, J. S., & et al. (2017). Tabiat Bersarang Dua Spesis Stingless Bee Koloni Di Pulau Ambon Muluku. *Seminar Nasional Perlebahan.* Bogor: IPB.

Michener, C. D. (1961). Observations on the Nests and Behavior of Trigona in Australia and New Guinea (Hymenoptera, Apidae). *American Museum Novitates.*

Michener, C. D. (2007). The Bees of the World. 2nd Edn. *The John Hopkins University Press.*

Nogueira-Neto, P. (1997). V_i_d_a_ _e_ _C_r_i_a_ção_ _d_e_ _A_b_e_l_h_a_s_ _I_n_d_íg_e_n_a_s_ _s_e_m_ _F_e_r_r_ão_._ _S_ão_ _P_a_u_l_o_:_ _E_d_i_t_o_r_a_ _N_o_g_u_e_i_r_a_p_i_s_ _

Rasmussen, C. (2008). Catalogue of the Indo-Malayan/Australasian stingless bees (Hymenoptera: Apidae: Meliponini). *Zootaxa 1935.*

Sakagami, S. F. (1975). Stingless Bees (Excl. *Tetragonula*) from Continental Southeast Asia in the Collection of Bernice P. Bishop Museum, Honolulu (Hymenoptera, Apidae). *Jour. Fac. Sci. Hokkaido Univ., Ser. VI, Zool. 20 (1),.*

Sakagami, S. F. (1978). Tetragonula Stingless Bees of the Continental Asia and Sri Lanka (Hymenoptera, Apidae). *Jour. Fac. Sci. Hokkaido Univ. Ser. VZ, 2001. 21(2).*

Sakagami, S. F., & Inoue, T. (1985). Taxonomic Notes on Three Bicolorous *Tetragonula* Stingless Bees in Southeast Asia. *Kontyii, Tokyo, 53 (1): 174-189.*

Sakagami, S. F., Inoue, T., & Salmah, S. (1990). Stingless Bees of Central Sumatra. *Sapporo; Hokkaido University Press.*

Sakagami, S. F., Inoue, T., Yamane, S., & salmah, S. (1989). Nests of the Myrmecophilous Stingless Bee, Trigona moorei: How do Bees Initiate Their Nest Within an Arboreal Ant Nest? *Biotropica, Vol. 21, No. 3.*

Schwarz, H. F. (1939). The Indo Malayan Species of Trigona. *Bulletin of AMNH.*

Starr, C. K., & Sakagami, S. F. (1987). An Extraordinary Concentration of Stingless Bee in the Philippines, with Notes on Nest Architecture (Hymenoptera: Apidae: *Trigona* spp). *Insectes Sociaux Vol 34.*

Welzen, P. C., J. F., & Alahuhta, J. (2005). Plant distribution patterns and plate tectonics in Malesia. *Biologiske Skrifter 55*, 199-217.

Wikipedia. (2018, Feb 17). *Dipterocarpaceae*.

Wikipedia. (2018, Feb 10). *Shorea*. Retrieved from https://en.wikipedia.org/wiki/Shorea

Wikipedia. (5 Desember 2018). *Daftar_kayu_di_Indonesia*. Retrieved from id.wikipedia.org: https://id.wikipedia.org/wiki/Daftar_kayu_di_Indonesia

Xiong, C. S. (2015). A Conservation Assessment of Stingless Bees (Apidae: Meliponini) in Singapore. *Cohort AY2014/2015 S1*.

Section 2 – Java, Madura & Bali

Meliponiculture tour and excursions in Java, Madura & Bali.

Contributors:

Wan Iryani Wan Ismail, Bajaree Chuttong, Norita Widya Pangestika, Mohd Razif Mamat & AHJ

Figure 84 Panoramic view enroute to Kintamani.

Contents in Section 2

Prelude to Section 2

I was told of a Javanese fable where in days of old among travellers in the past if one lingered at the mouth of the dragon, one has to traverse throughout its whole body and exit at the tail, else one will be trapped in and swallowed by the dragon for eternity. This fable I heard when I was travelling at Ujung Kulon (meaning the west end). I realized the island of Java resembles a serpentine dragon of sorts.

What I have been doing is the opposite direction of my travels in Java over the years. The sequence of my presence there from as far back as 1993 as I recall was East Java 1993 (Horticultural interests), West Java 1995 (Textiles and crafts interests), Central Java 1996 (Handicrafts and novelty), Bali 1997 (Garden Décor), East Java 1999 (Shell crafts and Malacology), Central Java 2000 (Woodcrafts and Furniture). In the new millennium, I was scouting aimlessly in the East, Central and Western parts of the island and looking to venture into stone crafts, glass crafts and interior décor. It was not until some major mishaps and natural disasters that I paused my Java travels for about ten years.

Some new interests developed and trended in Meliponines, and it is now honey production that makes me look again at the island with new vigour and enthusiasm. By 2014 when I produced my first book on Beescape and meliponine conservation, I started making plans to revisit the island of my late mother's origin (she was of Betawi descent). Although she was brought up in a Dutch boarding school in Holland, she still could relate some of her childhood memories in Kemayoran, West Java. I revisited the regions she mentioned where her grandfather had a tobacco farm. Guess what? Absolutely nothing whatsoever was as she described.

Back to the fable earlier, I looked up some mythical history and found some interesting myths. From those old stories, I discovered that if I look intently at the topographic island map of Java, I see a lying giant serpentine dragon coiling the mountains, and I was probably doomed into living in its belly.

I worked out an impression of the beast and retraced my journeys on meliponiculture expeditions that constituted part of my initial project on the Meliponiculture atlas of the Nusantara region.

Mythology

The Javanese myth of Tantu Pagelaran explains that Batara Guru (Shiva) ordered the god Brahma and

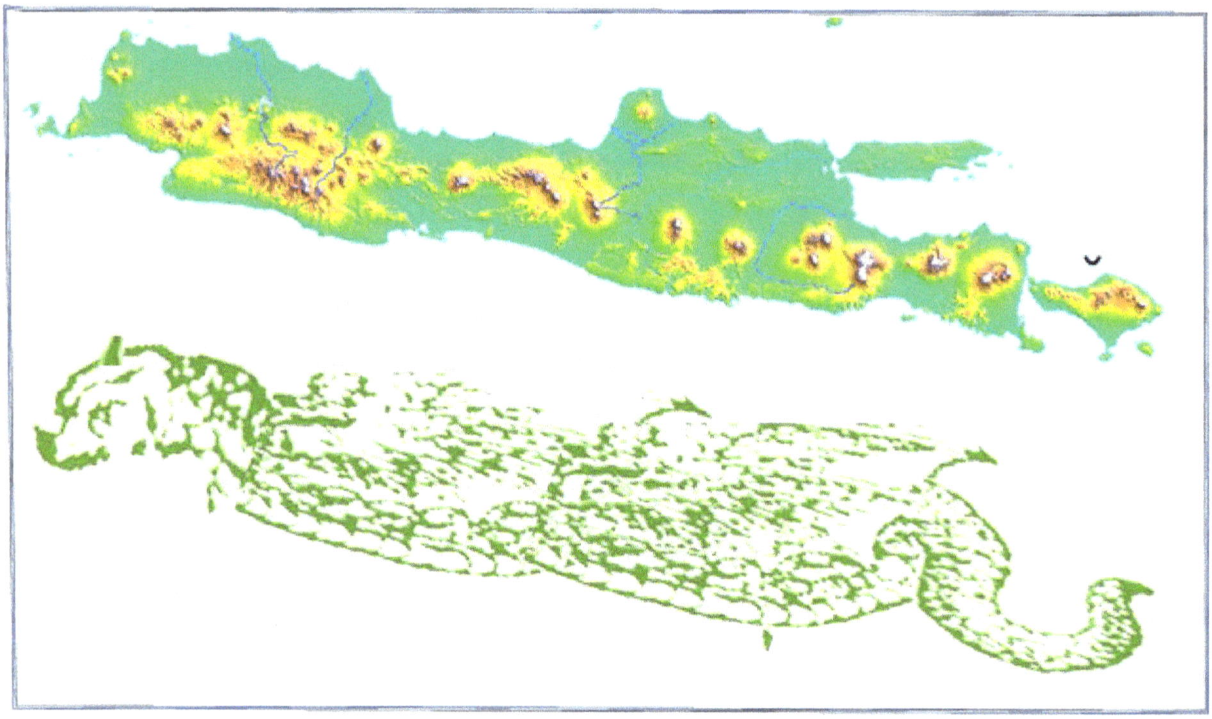

Figure 85 Naga Serpent of Java

Vishnu to fill the island of Java with human beings. However, Java Island was floating freely on the ocean, ever tumbling and always shaking. To make the island still, the gods decided to nail the island upon the earth by moving a part of Mahameru in Jambudvipa (India) and attaching it to Java.

The god Vishnu transformed into a giant turtle and carried a part of mount Meru upon his back, while the god Brahma transformed into a giant naga serpent and wrapped his body around the mountain and giant turtle's back so that the Meru Mountain could be transported safely to Java.

When Sang Hyang Shiva arrived at Java, he saw many *Jawawut* plants (the foxtail millet *Seraria italica*); thus, the island was named Java.

Meliponine Relationships with Jawawut.

Figure 86 Kinds of cereal Poster

Chapter 2-1

2-1.0 - Bogor, West Java

The Seminar Perlebahan Nasional in IPB Bogor was held on 19th August 2017. I attended a workshop on Urban Stingless Beekeeping and Honey Harvest in this seminar.

Figure 87 The Seminar Perlebahan Nasional in IPB Bogor was held on 19th August 2017

2-1.0.1 – Hambali Farm @ Baranangsiang, Bogor.

Later visited Baranangsiang in Bogor. Meliponary of Gregory Hambali, a botanist by profession.

Figure 88 Meliponary of Greg Hambali, a botanist by profession. Uses Baringtonia edulis *and assorted species of* Zalacca.

We visited Baranangsiang in Bogor. The meliponary belongs to Mr. Greg Hambali, a Botanist by profession, who is using *Baringtonia edulis* and various varieties of Salak. He Uses *B. edulis* and assorted species of *Zalacca* (Snakeskin fruit). As a botanist, he is proud to provide flowered florets and floral blooms in the Meliponary.

In addition to Salak palms, he planted *Medinilla speciosa* and *Manihot esculenta* (variegated cassava) to provide his stingless bees with the plant sap [for food or what? Construction material?] which is it,

Figure 89 Medinilla speciosa & Manihot esculenta

sap or resin, or gum?

2-1.0.2 - U3Bees @ Depok, West Java

I went to UI (University Indonesia) in Depok and visited U3Bees belonging to Dr Mohd Sahlan, Jeffry Pamungkas Lesmana & Andria Fitriada.

In University Indonesia (UI), there is a group of stingless bee breeders called 3bee (Trigona bee), consisting of 9 members, Mr. Jefry as founder, Pak Khodziq, Pak Taufik, Pak Andre, Pak Yoyo, Pak Yogi, Pak Ade, Mr. Nelson Mr. Sahlan who is one of the lecturers at the Faculty of Chemical Engineering UI. Pak Sahlan also engaged in Lebah propolis products such as honey, propolis, soap and cosmetics. For Lebah propolis livestock in the UI Forest itself, there are about 350 colonies in box hives. Allegedly there are several species of *Tetragonula*. UI Forest is suitable for the cultivation of Lebah propolis because of the supporting environment, and there are various types of food sources such as flowers Bride Tears, Calliandra flowers, Kuru-kuru, Kenikir (*Cosmos caudata*), Jengger Ayam (*Celosia* sp.), Sorghum & etc. Honey crops of this group have also been sold in Java and outside Java. For 1 litre of honey at a price of about 350 thousand rupiahs. In addition, the 3bee group also sells Lebah propolis colonies already in the wooden box at a price of around 150 thousand rupiahs.

Figure 90 to UI in Depok visited U3Bees belonging to Dr. Mohd Sahlan, Jeffry Pamungkas Lesmana & Andria Fitriada.

2-1.0.3 - Urban Bee Teuweul @ Sindang Barang, Bogor.

After the seminar on 20th Aug., we visited Dr. Mahani's Meliponary in Kavling Panorama Sindang Barang Bogor. Pak Andri, Farm Manager, in an interview on 4 September 2017, Pak Andri is Pak Mahani's trusted man who has been working for CV. Dr. Mahani is the director of CV. Natural Healthy Nutrima. Natural Healthy Nutrima has various honey and propolis products from Stingless bees (Lebah propolis) and a teacher at one State University in Bandung. His science background is Food Technology. Now he is taking his doctoral program at IPB and researching propolis. For the last few years, Dr. Mahani has been involved in the business industry that sells various Lebah propolis honey and propolis products and several other products. He also has a patent on how to extract propolis.

A large market demand makes buying raw materials of propolis and honey on a large scale a must. Luwu Sulawesi is one of Indonesia's largest honey-producing areas.

Figure 91 Urban Bee Teuweul Bogor

with *Wallacetrigona incisa*. For propolis raw material, he bought it from Java and outside Java. Meliponine honey product in the 250-gram bottle and propolis 6 ml extract per bottle is priced around 80-100 thousand rupiahs. In the last three years, he also began to collect Lebah propolis. There are about 30 colonies of the *Tetragonula* spp. and until now efforts continue to develop techniques for the propagation of colonies.

Innovation is impacted by the shape of the box that promotes the structure of the "vertical" interior nest. This system is difficult to harvest but to culture a minute species where the squeeze and sieve system is practised yields a rather satisfactory result.

Visitung Dr. Mahani's Bee Farm Urban Bee Teuweul Bogor.

(Malay Text by Dr. Wan Iryani Wan Ismail – UTM; Translated by Abu Hassan Jalil AKM)

Figure 92 A visual account of the visit by a group of Malaysian Genome Inst. Staff and a few individuals from Malaysia and Thailand.

A visual account of the visit by a group of Malaysian Genome Inst. Staff and a few individuals from Malaysia and Thailand.

Figure 93 visit to (kunjungan ke) Dr. Mahani's farm named Urban Bee Teuwul Bogor

Meliponiculture in Indonesia was started by obtaining the most suitable species of kelulut for commercialization. For example, the results of a 14th Asian Apicultural Association (AAA) conference visit in Jakarta, Indonesia, on 22-25 October 2018, a visit to Dr. Mahani named Urban Bee Teuwul Bogor was conducted. This farm is located in Bogor, Indonesia. They issued a comprehensive product that involved the Bogor community in the name of Nutrima, Sehat Alami since 2007. They are long-lasting and well developed because they use the knowledge of the most suitable type of meliponines to be commercialized in their fields, namely the meliponine species *Tetragonula* cf. *biroi.* This species was chosen because it is endemic in the Sulawesi region, which can produce more quantities of propolis and honey than other species, the best quality of propolis and honey in terms of antioxidants, suitable for farming in the city, easiest for breeding, disease resistance and the easiest to propagate colonies. This bee is also successfully kept because it takes the most preferred food sources, the main 3 being Mango, *Antigonon leptopus* and *Lagerstroemia indica,* recognized by the vernacular name Bungor. These three main types are chosen because they are sources of nectar, pollen and resin, which are prolonged for food and fostered nests. This plant also provides additional sources of nectar and resin. Other plants are also planted, like limes, bananas and Brazilian sunflowers.

The picture below explains: a) Urban Bee Teuwul Farm, Bogor, Indonesia, b) Marketing information through social integration by Dr. Mahani, c) *Tetragonula cf biroi* wrinkle nests, d) *T. biroi* eggs and honey bags, (e) *T. biroi*

Figure 94 Field trip to Bogor, Indonesia Source: Dr. Mahani, Ladang Urban Bee Teuwul Bogor

Figure 95 Nutrima Sihat Alami promotional Poster Source: Dr. Mahani, Ladang Urban Bee Teuwul Bogor, Indonesia

Photo Album from Dr. Bajaree Chuttong during Bogor trip.

Figure 96 Display cabinets @ Urban bee

Post-Harvest Equipment

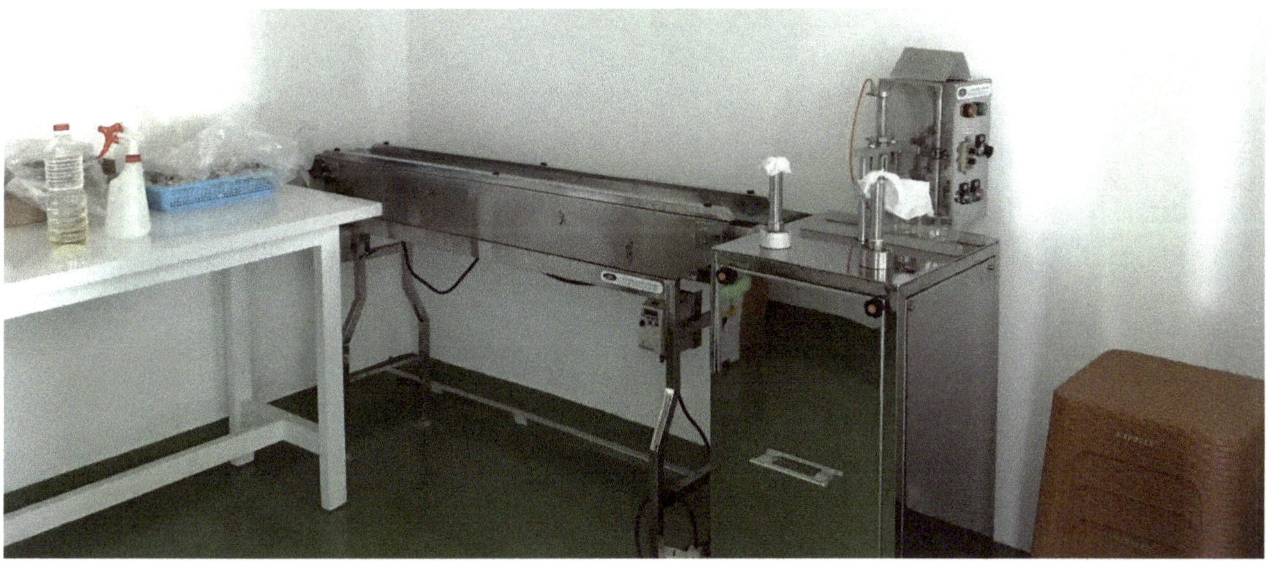

Figure 97 Post Harvest equipment

Figure 98 Harvesting and nest entrance closeup.

2-1.0.4 - Andre Bee Farm @ Cibereum, West Java

Our next destination was Cibeureum which is situated about 80 km south of Jakarta and about 20km (half hour from Bogor) *Tujuan kami berikutnya adalah Cibeureum yang terletak sekitar 80 km selatan Jakarta dan sekitar 20 km (setengah jam dari Bogor)*

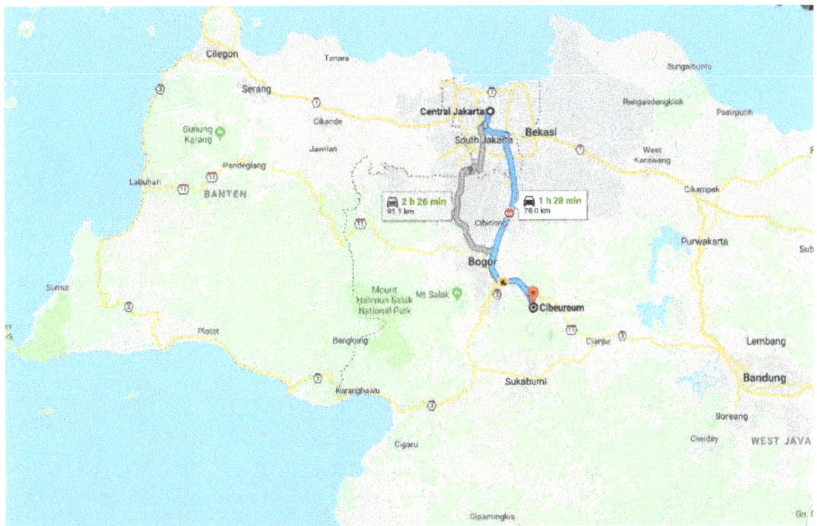

Map 6 Route to Cibereum from Jakarta Rute dari jakarta ke Cibereum

Figure 99 Two-mile hike to access the bee farm and a look at the Meliponary activities.

Figure 100 After traversing streams and some neighbouring farming plots for about 2 miles, we reached his bee Farm and Andre gave some dehydrated "Stingless Bee Pot-Pollen" to replenish my stamina

Figure 102 We were intrigued by his double box contraption.

Figure 101 we examined Andre's innovation in double box hive.

Photo Album from Dr. Bajaree Chuttong during Bogor trip

Figure 103 The inner box could be removed for easier excess to the honey pots, hence easier harvesting

Figure 104 Sieving harvested honey

We were treated to some home-cooked (by his lovely wife) village food back at his house for lunch. Overall, it was quite an exhausting trip.

Photo Album from Dr. Bajaree Chuttong during Bogor trip.

Figure 106 With the brood cells in the innner rectangle, the bees made the honey pots in the outer rectangle.

Figure 105 Nest entrances of Tetragonula *sp. in his box hives*

Figure 108 Along the way we passed a turnip farm and workers were washing newly harvested turnips.

Figure 107 The racks of his box hives are on an elevated ground above the turnip farm.

Chapter 2-2

2-2.0 - Banten, West Java

2-2.0.1 ASEAN Stingless Bee Workshop 2019

It was a fortunate coincidence that I was invited to present at an international workshop for the Stingless bee in the ASEAN region. The venue was the Int Convention Exhibition (ICE) in Bumi Serpong Damai (BSD or Bintaro Sonoan Dikit) city in Tanggerang, Banten, West Java. Fortunate because I needed to satisfy my much-anticipated exploration of the region of West Java on the meliponiculture practices. And so, the plan was devised.

Map 7 The excursion route was ultimately developed to search Heterotrigona itama endemic to Java.

I gave all during the workshop as I met many colleagues in the same field. Also renewed many old acquaintances and made quite a few new friends. I made plans to tag along with Dr. Sih Kahono of LIPI, who was scheduled to travel to Ujung Kulon (West End) the next day. We opted for some local stalls nearby for some authentic Javanese cuisine in place of the costly hotel dinner.

I pulled Yusuf Efendi of Dumai, who agreed to accompany me, and we tagged with Dr Sih Kahono, who had a guest, Mr. Yoshei Kambayashi, with him. After the afternoon, we set out from Tanggerang and headed for Pandeglang, Southwest of Jakarta.

Figure 109 Feeling a little woozy, I gave my summation sitting down. Recieving a token after the Workshop.

Along the way, we stopped for lunch and a spot of Durians. It was a cosy place with 'pangkins' to sit on for the traditional way of having meals without chairs. We then arrived at the first destination, a house cum store for the Cooperatives stock of wild *Apis dorsata* honey (locally referred to as 'Madu Tawon Gung' or 'Madu Sialang'. Sialang is the tree whose branches these bees usually set their nests on.

2-2.1 - Pandeglang Regency: Ujung Kulon

Travelling for the next 4 hours took its toll. We were deadbeat because the last hour was on an untarmacked road. It was strewn with crusher run and quite dusty, as it was the dry season. We arrived at Teluk Paraja and lodged the night at Penginapan Sunda Jaya in Taman Jaya, Sumur, Pandeglang Regency. This place is the westernmost accommodation for the Ujung Kulon National Park.

We had a good night's rest and headed for the village we intended to visit. This was to view the cooperative's Meliponary in Kampung Pete (local pronunciation of Petai - *Parkia speciosa*). My ultimate plan was to observe the endemic Javanese *Heterotrigona itama* that was rumoured to occur in this region.

When we arrived at the village, I was fascinated by the many varied motifs of the woven bamboo screen walls of the houses. I reminded Yusuf to capture as many different motifs he could find. The village is very peaceful and quiet. There was practically no one around in the first few hundred yards, and it was already 9 a.m.

The houses, though rustic and spartan at a glance, had many had satellite dishes. The honey business must be quite lucrative here. I guessed the occupants must be watching some exciting TV show. We had to park at one of the houses when we arrived at a small unpassable bridge over a ditch or stream. Right after the bridge at the stream's banks was the Meliponary.

Figure 110 The Meliponary was like a long shack about 30 meters long with 2 shelves and a hundred or more hive boxes packed one on top of another.

The Meliponary was like a long shack of about 30 meters with two shelves and a hundred or more hive boxes stacked one on top of another. Although well shaded with coconut leaf attap, there were boxes of failed hives at one corner. Overcrowding was an obvious factor. Yoshie Kamabyashi (red shirt) immediately started capturing images while I explained to Dr. Sih Kahono (in black T-shirt) that the overcrowding was causing drifting while we observed some swarming.

The other *Tetragonula* boxes seem to have the entrance hole too large, so the bees modified their entrance.

Figure 111 The other Tetragonula *boxes seem to have the entrance hole too large so much so the bees modified their own entrance.*

Figure 112 There amongst the stacks of shoebox size boxes, was one odd shaped hexagonal box of what appeared to be the Heterotrigona itama *colony. Others were of* Tetragonula.

It was exhilarating to observe the *H. itama* endemic to Java in this west end. A pre-war paper indicated *H. itama* occurrence in Pelabuhan Ratu, which is 5 h away in southernmost Jakarta. This indication was a myth among Meliponiculturists in other parts of Java as most of them 'import' this popular species from south Sumatra. This location was my next destination.

I interviewed one of the co-op members (Pak Andak), who personally owned a genuine *Heterotirgona itama* colony that he gathered in the nearby forest. He mentioned that it was a rare occasion for him to discover this rather large bee instead of others (*Tetragonula*) during one of his colony hunting sprees. For that, he decided to keep this one at the back of his house, away from others. This type he says that the locals call it Teuwuel Ga'ak, meaning crow as it was as black as a crow.

I later rested at the front of his house while his family served us coffee and some honey. They were anxious to know what I thought of the taste, and I could tell that the honey was a blend of a couple of

Figure 113 I interviewed one of the coop members who personally owned a genuine Heterotrigona itama colony that he gathered in the nearby forest.

species of *Tetragonula*. From what I deduced of the species they had, it was most likely blended honey from *T. fuscobalteata* and *T. laeviceps*.

Chapter 2-3

2-3.0 - Sukabumi Regency West Central Java

2-3.0.1- Pelabuhan Ratu

We arrived at this destination in a little more than 5 h, and I ventured into one of the locations of a forest reserve to try to source some wild feral colony. I even stopped at a roadside stall selling dried

Figure 114 I ventured into one of the locations of a forest reserve to try to source some feral colony. I even stopped at a roadside stall selling dried salted fish in the hope I might see a bee or to inspire my quest.

salted fish (as we know stingless bees love to forage salts) in the hope I might see a bee or inspire my quest. As luck would have it, nothing at all was close, that I could encounter.

With our energy fading, we headed for a hotel nearby and I thought to again tomorrow morning. In the meantime, I took advantage of the seaside setting to enjoy a good seafood dinner.

Having rested from the strenuous journey and frustration at not finding any bees, we set out to Sukabumi city.

2-3.0.2 – Mangkalaya, Sukabumi.

We arrived at Mangkalaya village in Sukabumi at the house of Fadil Anwar. As it appears, Mr. Fadil is an avid hobbyist who keeps bees and dabbles with some peculiar breeds of rabbits. His bees are kept all around his house compound, pleasantly gardened. He welcomed us with some coffee and local breakfast tidbits and snacks. Below are some photos of his colonies.

Figure 115 Top photos shows the H. itama *hive he bought from Lampung, Sumatra; Middle photo shows a failed imported* T. biroi *colony that lost out to a more aggressive local* T. laeviceps *beside it.*

2-3.0.3 – Nalbinding, Sukabumi.

Moving ahead, we reached Nalbinding, where a Meliponary was managed in a religious school turned into an eco-tourism establishment. It is run by Iping Madu of Pondok Pesantren Yatim. The place was bustling with visiting children on school excursions. It was the school holiday season, I suppose. They had horse rides, Merino sheep on display and numerous agricultural activities. One of which was stingless beekeeping, the one I came for.

Figure 116 The Meliponary consisted of 10 T. biroi *colonies imported from Sulawesi and stands in the original vertical box hives they came in, on the upper shelves were local* T. laeviceps *shoebox type wooden hives.*

The Meliponary (elongated) is shaded by Pine trees (*Pinus mercusii*) on one side and faces a plot of floral beds (*Portulaca* sp.) on the other side. Pak Iping explained that the meliponary is fairly new and requested some advice. I first indicated that the upper shelves had the boxes too close to each other and would face much confusion when the swarming seasons came. The vertical boxes may pose some harvesting issues if harvested in the same vicinity as the shoebox types have different harvesting techniques. These may have battling between the *T. biroi* and *T. laeviceps.*

The foraging paths were unclear because the hives facing the pine simulacra will be diverting to the floral side, and much crisscrossing will prove inefficient as foraging flight paths.

Figure 118 The Meliponary (elongated) is shaded by Pine trees (Pinus mercusii) on one side

Figure 117 Box hive racks among Pine trees

2-3.1. – Bandung Regency

We spent the night in Bandung at a hotel with some antique décor. The Paintings at the stairway intrigued me as they depicted the daily life of Bandung. Pak Teguh soon arrived and brought us to Pak Romli's, who later took us out to dinner at a restaurant.

2-3.1.1 - Banjaran Wetan, Soreang

Figure 119 From left Enjang's cousin, Enjang Yana, Abu Hassan, Amir the shopkeeper, Yusuf Efendi.

That morning we made our way with a hired car to a location indicated by Nzank Yana. We searched for Banjaran Soreang Kp. Babakan Sadang Desa Banjaran Wetan Rt 01 Rw17. We found Amir, a shopkeeper, at the location. Amir (2nd from right in the photo in front of his shop) said he knew the name (pronounced 'Enjang') and could bring us to the cousin's house.

The cousin acknowledged us and took us to his Meliponary a short distance away. His place was idyllic though a little bit tight for

Figure 120 The place was idyllic though a little bit tight for the total hives and the varied vegetation he had

the total hives and the varied vegetation he had there. The owner is very enthusiastic because we could see that he experimented with many types of containers and hive materials and the vegetation included almost every flower he could find.

Figure 121 The owner is very enthusiastic because we could see that he experimented with many types of containers and hive materials

Figure 122 Bamboo hives abound and some in clay pots and one neatly constructed traditionally modern scaled model house.

Figure 123 The owner expressed his dreams of getting a larger plot to place his hives and plants.

We could not spend much time there because we had to get back to Bandung to check out. Furthermore, we were scheduled to meet up with Teguh to proceed to Purwokerto for our next plan: to meet up with the Dean of the Biology Faculty of University Soedirman for an official signing of a collaborative MOU.

Chapter 2-4

2-2.0 - Nusa Kambangan Island - Cilacap Regency

Arriving in this Province reflects on the recent month's course held here at Prawita Garden in Ajibarang.

Figure 124 The last trip was when we had a course for some participants in Prawita Garden and then ventured to Nusa Kambangan Island.

The last trip was when we had a course for some participants in Prawita Garden and then ventured to Nusa Kambangan Island. Still fresh in memory was certainly quite a trip on that island. Kristosie Imanuel did a follow-up trip

2-2.0.1 - Nusa Kambangan

The first trip to Nusa Kambangan was on a quest for *Lepidotrigona javanica.* Prompted by old records of sightings of the species almost 128 years ago by Gribodo. "Lepidotrigona javanica (Gribodo, 1891) Type locality: INDONESIA "Giava" (=Java) (1 worker); Dalla Torre 1896: 579 (citation); Friese 1914a: 13 (distribution); INDONESIA, Nusa Kambangan)" excerpt from Rasmussen's catalogue 2008. With that, we made a trip to Teluk Penyu in Cilacap, to a fishing village to hire a boat to take us to Nusa Kambangan Island.

The boat ride was short, and we landed on Tambakreja, a stony barrier of the island (there was no sandy shore or jetty to dock), just a rock outcrop that we had to climb up. Although there was a footpath, the loose cobble track was not an easy uphill trek for me at my age. The guys, Pak Didi (the guide), Mistem and Teguh (from Prawita Garden, Ajibarang), Yusuf Effendi from Dumai (who was accompanying me on the trip), Kristosie Imanuel from Purbalingga (a student who just finished his practical stint in Ajibarang), they were very supportive and patient of my predicament.

We started trekking uphill, passing some limestone rock outcrop. We found several stingless bee nest entrances at the rockface sides. We found a few more nest entrances, and then we decided to place some bait dishes as a few traps. The guys collected specimens, and then we continued till we came to some old Dutch colonial building ruins. There we stopped and rested a bit while exploring the ruins.

Trekking in Nusa Kambangan, Centra Java.

Figure 125 Trekking @ Nusa Kambangan Island

Figure 126 Found some interesting nest entrances.

I decided we needed a lab to examine the specimens microscopically after taking specimens. They decided that the closest was University Soedirman in Purwokerto. That is where Kristosie was studying Biology. I asked him to make an appointment to meet his Dean of the Faculty. Kristosie (the student working on this project) did some follow-up visits and images to the island.

Figure 128 Kristosie reached the lake which we opted earlier to go to the beach instead

Figure 127 Pathway to the lake and a couple of peculiar nest entrances.

With an appointment set for the next day, we gathered the necessary materials, left the island, and proceeded back to Ajibarang. But before that, I had to conduct a 2-day course on Meliponiculture in Ajibarang upon our return, as promised.

Figure 129 Photo gallery of Nusa Kambangan expedition Sept 2019 through the dry creek up to the Kali Kencana beach.

Figure 130 The Lepidotrigona cf javanica found previously on the island by Pak Teguh of Prawita garden.

Chapter 2–5

2-5.0 - Central Java

2-5.0.1 - Pak Teguh @ Banyumas.

A look at **Pak Teguh** (see photo below) of Banyumas in his Beefarm in Purwokerto, Central Java. Met Pak Teguh at the AAA 2018 convention and managed to obtain some of his pictures of his activities, type of stingless bees and records of the vegetation in his Bee Farm.

Pink guava cultivated in his farm was started approximately from 2 y ago. Initially, the fruiting was quite inconsistent, but now, the yield looks promising with more bees.

Figure 131 Pink Guava cultivated at pak Teguh's farm.

His farm also had *Calliandra calothyrus,* commonly found in Java, Indonesia.

Figure 132 Photos of Pak Teguh's Meliponiculture. Foto Budidaya Klanceng Pak Teguh

Some trees grow wild in the vicinity of his farm. Flowers of the Sapindaceae family seem to attract the bees with their abundant pollen. Some monkeys frequent the trees, a very tall tree, and flowers fall off after they bloom. The bees still forage the flowers that have fallen on the ground, apparently having still a lot of pollen and nectar. Looks like the *Pometia pinnata* or the "Fijian longan". This tree is native to PNG and is also found on Sulawesi. *Pometia pinnata* is a large tropical hardwood and fruit tree species, with common names including matoa, taun tree, island lychee, tava, Pacific lychee of the plant family Sapindaceae. The wood is known as Kasai in Malay.

"These are several samples of kelulut besides *laeviceps* in Java that have not been identified because I did not dare to cut trees. Being tried with traps and seeking for identification," he said.

Figure 133 Wild flowers attract monkeys and the bees with their abundant nectar and pollen

Map 8 Location of Pak Teguh's Farm Purwokerto in Banyumas, Central java.

A map of his location is attached, and some more closeups of his varied *Tetragonula* spp. nest entrances at his location.

Figure 134 Some wild nests at Pak Teguh's location

Figure 136 Some unidentified Tetragonula *nest entrances.*

Figure 135 Calliandra calothyrus *commonly found in Java, Indonesia*

2-5.0.2 Prawita Garden, Ajibarang, Purwokerto

Figure 137 The 2-day course on Meliponiculture held in Prawita Garden

Prawita garden in Ajibarang village is a model Bee Garden with fruits galore and floral vegetation. Both honeybees and stingless bees are kept in about 25 locations spread among the community. Besides grapes, Pak Teguh also attempts at producing strawberries and pink guava, which is coming out quite successfully.

Figure 138 Pollination of grapes by stingless bees

The next visit was to the garden itself and some Meliponiculture spots where Teguh has placed his hives.

Figure 139 We found boxes of different shapes sizes for different species and all different techniques. Possibly part of a frand experiment.

We also met someone culturing *Tetragonula drescheri* in Purwokerto, Banyumas District. This experience made me recall a particular reference.

(Ref: **Type locality: INDONESIA** "M. JAVA - South Bandjarmas, Koebangkangkoeng, 25 meters, July 1935 (F. C. Drescher)" (1 worker, holotype); "E.JAVA. -Mt. Goemitir, Nov. 3 (R. van der Veen)" (1 worker, paratype); (Rasmussen 2008)).

Figure 140 Prawita Garden

Figure 141 Another species in yet another location.

Figure 142 Harvesting Tetragonula *honey at one of the locations allocated by Pak Teguh*

Chapter 2-6

2-6.0 - Faculty of Biology, University Soedirman (UNSOED), Purwokerto

Somehow the following day, I ended up in the Faculty of Biology, University Soedirman, Purwokerto. We had an initial meeting with the dean Hj. Imam.

Figure 143 The initial meet with Faculty of Biology, UNSOED

We agreed that we would be allowed exclusive usage of the Biology Lab facilities, and their students will be included in the research team for the Bee identification of specimens collected in Nusa Kambangan Island.

This expedition was followed up one month later with the signing of an MOU between UNSOED and AKM.

Done with the formalities, we quickly headed to Purbalingga to save daylight hours.

Figure 144 We were allowed exclusive use of the Lab facilities after signing of the MOU.

c

Chapter 2-7

2-7.0 - Purbalingga Regency

2-7.1 - Desa Panusupan, Purbalingga

The undulating high grounds and valleys of Purbalingga that hunters Fuad & Tusmin hunt for rare orchids, ornate plants and beehives as part of their daily living. We visited their homes to gather some stories of their past escapades. More importantly, we desired information on the occurrence of *H. itama*

Figure 145 The undulating high grounds and valleys of Purbalingga

(which was reportedly quite rare in Java). They explained that the feral hives they found existed on tall trees, and they had to climb to about 5 meters up to access. The trees in question were about 15 meters above a rocky stream (aptly called Sg. Deng).

The *H. itama* hives were found 5 meters up the tress then transferred into bamboo after retrieval for ease of movement. We later visited their houses in the village to view the progress of their finds.

Along the way to Purbalingga, we were entertained by some street performers at a traffic lights junction. Unfortunately, it was dark by the time we arrived, but we had our trusted torchlights to light the way and our inspection. The first stop in Panusupan Village was Fuad's home. There we met with Tusmin, who brought us to his home.

Figure 146 They had to climb to about 5 meters up to access. The trees in question were about 15 meters above a rocky stream (aptly called Sg. Deng). Fuad & Tusmin up the tress - circled red

At Fuad's house, we saw some orchids and ornamental plants he turned in Bonsai. As it turns out, Fuad is quite a nature lover and interested in horticulture.

Figure 148 The H. itama *hives found 5 meters up the tress then transferred into bamboo after retrieval for ease of movement.*

Figure 147 Some hives ibc H itama colonies at Tusmin's house

Chapter 2-8

2-8.0 - Special Regency of Jogjakarta.

2-8.0.1 - Jogjakarta.

We now come to Jogjakarta, the Javanese centre of the Arts and higher education. I have to recollect the times I spent here roaming the countryside and neighbouring towns. My last visit here was to attend the National seminar on Bees at the University of Gajah Mada (UGM) and an expo in Ambarrukmo Hotel a month ago. Coming back to my analogy of the giant serpent Naga, Jogjakarta would be the innards of this beast.

Figure 149 Although the conference in Ambarrukmo Hotel was mainly on ruminants and husbandry the exhibition stalls were open to other agricultural practices including Meliponiculture. The response on our booth was quite overwhelming with all our literature bought up by a UGM Dean. (Top right photo)

Although the conference in Ambarrukmo Hotel was mainly on ruminants and farming, the exhibition stalls were open to other agricultural practices, including Meliponiculture. The response on our booth was quite overwhelming, with all our literature bought up by a UGM Dean.

Figure 150 After the exhibition was the Indonesian National Seminar on Tropical Bees. It was a time of jubilation to meet old friends in the same field.

Even further back reminiscing, I was in Jogja in August 2017, when I visited some beekeeping groups in Central Java, the Magelang Community Beekeeper group comprising 25 members and the Klaten Community Beekeepers Group comprising a dozen members. These groups were also under the advisement of various higher learning institutions.

2-8.0.2 – Haris Muhammad's Farm @ Magelang, Central Java

I arrived in Jogjakarta at 6 a.m. by the night train from West Java and headed to Magelang town about 50 km Northeast of Jogja. It was nearly 8 a.m. when we met up with Wiwied (a trusted friend of more than 10 years). Arrived Magelang at about 10 a.m. in Desa Kebon Rejo, Kecamatan Candi Mulyo Kabupaten Magelang.

Haris Muhammad is a stingless beekeeper's group pioneer in Kebon Rejo, Magelang area. Personal experience that makes Haris start raising 'Lebah propolis' (the vernacular name for stingless bee, although now to be known as Lebah Propolis) was from the accident that caused his wife to pass away, and Haris himself suffered a broken leg, right foot, and leg bone could be seen from the outside because of the compound fracture. After two years of medical treatment, he stopped it, because there was no satisfactory result. Then the doctor suggested that Haris rub his legs with Lebah propolis / Lanceng honey because of the higher antibiotic content than honey from *Apis* bees or known to the locals as 'madu Tawon'. After 3.5 months, his feet healed, and the wound closed. Originating from his own experience, he felt that lanceng honey had tremendous benefits. Honey from lanceng bees is sold at 60,000 Rp per 150 ml. In addition, they also sell bee pollen that has been packaged in capsule form. Finally, in 2010, Haris and his friends, including about 25 people, began to breed the stingless bees.

2-8.0.3 - Widodo @ Klaten, Central Java

About 2 p.m. we arrived at Desa Padangan Rt.02 Rw.03, Glodogan, Klaten Selatan, Klaten on 21 August 2017, at Pak Widodo's farm, mostly developed from bamboo hives. Not much different from the experience of Pak Haris from Magelang, Pak Widodo from Klaten started raising Lebah propolis bees when his wife suffered kidney failure and was suggested by his wife's brother to consume honey Lebah propolis. This advice came about because of the experience of some people who recovered from several diseases such as Hepatitis C and Diabetes. Pak Widodo himself began raising Lebah propolis bees in May 2014. Raising Lebah propolis bees brought many obstacles earlier and at the start, such as the number of colonies that died or disappeared. But from the failure, Pak Widodo learned many things. For beginner beekeepers, Pak Widodo suggested breeding the species *Tetragonula laeviceps* because this species is endemic in Java, easier to obtain, and more easily finds its food (flowers). First time rearing, Pak Widodo used the stingless bees from various *Heterotrigona* species in about 500 colonies, but many colonies failed. However, Widodo's turnover now reaches 20 million Rp / month due to his tenacity and patience. In addition to selling honey, Pak Widodo also sells colonies in bamboo hives and wooden boxes to maintain colonies in various areas inside and outside Java. Widodo's

Figure 151 Packaging using cotton jersey material for transportation of hives. Old fridge pump for large species

breeding system with his friends is to put some bee colonies of the *Tetragonula* sp. type. And *H. itama* spread over several locations to minimize the competition for the source of food. There was a spread of about ten colonies in logs or wooden boxes from each location. He receives bamboo hives from local hive hunters. The material to transport hives is packaged using a cotton jersey (see Figure 154). Old fridge pump for large species (e.g. *H. itama* from Sumatra), and smaller species like *Tetragonula* sp. he uses cactus flowers *Epiphyllum strictum*. His honey harvesting method is a squeeze and sieve method.

Figure 152 Left: Uses cactus flowers Epiphyllum strictum; *Middle: Receives bamboo hives from local hive hunters; Right: Packaging using cotton jersey material for transportation of hives*

2-8.0.4 – Sugeng Apriyanto's Farm in Gunung Kidul, East Central Java

We headed to Dusun Ngrandu, Desa Katongan, Kec. Nglipar Kab. Gunung Kidul Pak Sugeng's farm of 400+ hives in terracotta. They use the Musa "1000 Fingers" cultivar**,** Pak Sugeng Apriyanto is one of the Lebah propolis beekeepers who show us their stock of bees, about 11 years ago they started breeding stingless bees (Lanceng), and members of his group now include 30 people. Some of the products sold are Lanceng honey, bee pollen, propolis, and royal jelly. The price of honey sold 220,000 Rp / 500 ml and propolis 120,000 Rp for 10 ml. In addition, this livestock group also opened a training

Figure 153 Calliandra calypthropis, Antigonon leptopus & *Musa '1000 Fingers' cultivar are part of the foraging sources for the bees.*

centre and beekeeping cultivation of Lanceng with nominal fees and one training reaching 1.5 million. Pak Sugeng and his group also received support from the Yogyakarta Provincial Office of Estate Crops (DIY). What is unique from Pak Sugeng's farming style compared to other Lanceng livestock groups is using a clay vessel used as a maintenance place. Now Pak Sugeng has thousands of scattered colonies consisting of 1950 colonies in clay vessels and 650 colonies in bamboo. Honey harvesting is done once a week from the alleged colonies of the type *Tetragonula* sp. Some of the plants supplied for the bees are *Calliandra* flowers, Bride Tears, and Bananas of the Musa '1000 Fingers' cultivar.

Calliandra calyothropis, Antigonon leptopus & *Musa* '1000 Fingers' cultivar is part of the foraging

Figure 154 Lanceng Gunung Kidul, Central java. Meliponiculture in Bamboo and Clay-pots

sources for the bees.

Chapter 2-9

2-9.0 - North Central Java

2-9.0.1 - Sobo Wono Bee Farmers at the Pati District.

Honey And Klanceng Bee Farmers "Sobo Wono" Binaan Perum Perhutani,

by Norita Widya Pangestika.

Norita started her portion of the trip from her hometown in Kudus and headed for Pati District. Pak Jafar Shodiq began raising bees in 1990, located in the village of Lahar, Tlogowungu, Pati Subdistrict; bee colonies were cultivated in the yard around the house and the house of his son, which was not far from his residence. There are around 4000 klanceng bee colonies in Bamboo and wooden boxes and hundreds of 'imported' *Apis mellifera* bee colonies. For bees, *A. mellifera* is cultivated routinely and "off and on" in several places and harvested several times a year. Pak Jafar also received assistance from Perhutani (Govt Forestry Dept) for the process of harvesting A. mellifera honey. In comparison, klanceng bees are harvested only once a year. The number of klanceng bee colonies that reached thousands made the yield of Pak Jafar's meliponine honey reach 300 kg in one harvest, with a selling price of 460 ml/bottle for Rp 135,000, while *A. mellifera* honey sold for Rp 75,000 with the same bottle size.

Figure 155 The main shack housing box hives and bamboo hives

Pak Jafar got the Klanceng bee colonies by buying from the community, which had a sign or box. Pak Jaafar also moved colonies from bamboo to vessels or pots made from clay, but the bees fled to leave the vessel (probably not compatible due to the heat), so Pak Jafar prefers colonies to be cultivated in bamboo that has a large diameter, different from how they breed in Gunung Kidul using clay vessels as klanceng terracotta hives. These hiving needs to be further investigated by differences in behaviour, perhaps because of differences in environment, temperature or different types of bees that make these bees also have different behaviours in choosing a host for their nesting place. Pak Jaafar's beekeeping is the main livelihood to support his family.

Due to his age, Pak Jafar, who has begun the "SOBO WONO" bee cultivation, is now managed by Pak Jaafar's son and Pak Hasan, who is Mr. Jaafar's son-in-law. With that, his bee farm can now be said to be a family business.

Pak Jaafar also often visits students, lecturers, and researchers from home and abroad, including British and Australian enthusiasts. In addition, several times he had received a request to export his honey to Britain and Australia. The availability of natural food sources from the gardens around the house made Pak Jaafar's bees happy to stay. Some of the plants around the house and the yard of Jaafar were 'Randu' trees, Rambutan, some Mango trees, lots of Coconut trees, Star fruit, Banana plants, Tangerine

Figure 156 Top left: Norita with Pak Jaafar; Bottom left: Pak Jaafar inspecting his hive racks.

oranges and several types of flowering shrubs. But there are still dozens of colonies that were discovered to have absconded/fled, died off, or moved and only realized when checking at harvesting time. This late discovery is usually discovered because the honey harvesting is done once a year. This frequency minimizes colonies that disappear from their nests due to overwhelming human activities.

Besides that, because klanceng bee honey production is minimal, Pak Jaafar must be patient to harvest only once a year.

Figure 157 Left: Closeup of bamboo hives stacked horizontally; Right: The Sobo Wono label.

Figure 158 Top & Bottom left: Bamboo hives hung vertical, horizontal and diagonal; Top & Bottom right: Rack of clay or terracotta hives.

Bamboo hives hung vertical, horizontal, and even diagonal, yet they all seem to provide compatible nesting cavities in the bamboo internodes. However, the terracotta hives are incompatible as some have failed, probably due to overcrowding. Further studies need to be done here.

The meliponines in his farm have not been identified, and there may be a few different species, but all are harvested as one, i.e. *Tetragonula* sp. It is difficult for the problem of species identification, especially if some of the *Tetragonula* types, even among scientists, in Indonesia alone, there are also no taxonomists or experts who specialize only in the identification kelulut. When I met Dr. David Roubik in Malaysia, I asked whether the bee species could be known from the funnel? Then he said

Figure 159 The shady surrounding of the Meliponary

almost but not always. I conclude if you want to know that the bee species must ask confirmation from a taxonomist specializing in kelulut. Because just by looking at the funnel or the type of brood cluster or layered if not stacked can be fully used as a benchmark. Hopefully, someday we will have taxonomists in the field of kelulut.

Chapter 2-10

2-10.0 - Eastern Central Java

Map 9 East Java & Bali Excursion Routes

We (Dr. Bajaree Chuttong & AHJ) started the East Java Tour by taking a flight from Jakarta (i.e. after the AAA 2018 Conference) to Solo City, better known as Surakarta in Central java. We were fetched at the airport by "Terry" (real name Tri Yulianto), who then brought us to meet Ms. Norita Widya Pangestika (our local Javanese translator and co-excursion member) who hails from Kudus city (North Central Java). We fetched her at the Surakarta Train station and proceeded to Terry's home and farm in Wonogiri (see map above).

Figure 160 Vegetation around Terry's home / farm.

2-10.0.1 - Tri Yulianto @ Wonogiri, East Central Java.

Home and Farm of Pak Tri Yulianto aka "Terry" (Desa Jendi, Kec. Girimarto, Kab Wonogiri, Central Java) (Indonesian Text by Norita Widya Pangestika. Translated by Abu Hassan Jalil)

Pak Tri is a hydroponic celery farmer who has begun to explore Klanceng bees. For about ten months, Pak Tri has learned to breed Klanceng and has 100 Klanceng colonies in the box. The main income is from Hydroponic Celery farmers, in a day can produce 10 kg of celery for 7000-25000 per kilogram. Besides celery, he also produces mint. We saw that he grew some cocoa and also coconut trees.

He rears a few oxen, and he found them with stingless bees inhabiting shed posts and rafters (made of bamboo). He started trapping them into cultivation boxes and has collected quite a number which prompted him to build more boxes and house them on a few racks around his farm. The honey he produced became quite a demand and then he sought a more consistent demand source than his local community.

As a result, in addition to selling hydroponic agricultural products, Pak Tri also sells the klanceng honey he gets from Sulawesi farmers.

He can sell 20-30 litres for 400-450 thousand rupiah per litre every month. Pak Tri admitted that he could not get the maximum out of raising klanceng bees because the source of vegetation around his house was still limited. This man has a basic agricultural science education background and conducts hydroponics courses. This activity was also supported by the service and friends of HKTI farmers. (HKTI = Himpunan Kerukunan Tani Indonesia). Translated, it means a localized Farmers Association.

The day after we left, Terry told us that he gathered his HTKI members and gave a briefing of what was discussed with us during our visit to his place.

All in all, he was elated by our visit and very much inspired.

Figure 161 Racks and at the Oxen shed

Figure 162 Racks and stingless bee hives

Photo Album from Dr. Bajaree Chuttong during the East Java trip.

Figure 163 Top left: Cocoa pods. Top right: Calliandra saplings; Bottom left: collection of traditional bamboo weaves; Bottom Right: stock lumber being sun dried.

2-10.1 Pak Harry Mustika @ Desa Sumurup, Kec. Bendungan Kab. Trenggalek, East Java.

Figure 165 Top left: Harry's Honey Brand Top Right: Group photo with Harry; Bottom left: Live fence of Antigonon; Bottom right: Harry's Cow Shed.

Figure 164 Having tea in his house: Rack of box hives at roof eaves

Pak Harry Mustika (see photo above 2nd from right). We now arrive at East Java Proper, i.e., the Regency of Trenggalek. We find that many here are active in agriculture, and most coffee growing and oxen rearing.

The bees in these regions seem compatible with oxen rearing and easily take up home in their shed pillars and rafters.

Figure 167 Tetragonula cf laeviceps internal nest structure.

Figure 166 Tetragonula *cf* biroi *internal nest structure.*

Above are some internal structures of trapped colonies cultivated in wooden boxes. These, as can be seen, are *Tetragonula* spp. Commonly found in Java.

However, these two images are the nest structure of *Tetragonula* cf. *biroi*, imported from Sulawesi, which clearly shows the development of a spiral brood arrangement. Mr. Harry has been breeding Klanceng for the past two years. He has 40 klanceng colonies taken from forests (bamboo internodes), people's homes, and limestone pits.

Difficulties when raising klanceng bees occur during the long dry season, so there is a lack of foraging, and the colonies are failing or dying. The vegetation found around Pak Harry's house is coffee trees, *Antigonon leptopus*, local sunflowers, Mexican sunflowers, Fire spike, Javanese Ginseng, *Calliandra*, Cassava, Rubber, Mahogany (not Swietennia- from S. America?), it could be other so-called 'mahogany', but only 2 Neotropical species are mahogany…), coconut, durian etc. For the price of honey from livestock, it sells for 175 thousand per 250 ml. Mr. Harry was also active in the land investigation (geological drilling). If he was working outside the island of Java, he also studied the klanceng that he encountered in the wild. Besides Mr. Harry, Pak Wito, a member of a livestock group has five Klanceng colonies that help pollinate his coffee land of 2 ha. From the results of coffee plantations, Pak Wito can produce 2 kg of robusta coffee per year. In addition, there is Pak Dwi, who has 15 colonies. His house is 500 m from Mr. Harry's house.

Local bees foraging *Begonia* flowers, and apparently, may forage some orchids.

Figure 168 Stingless Bee Honey & Orchids

2-10.2 - Pak Zaini @ Desa Ringinpitu, Kec. Kedungwaru Kab. Tulungagung, East Java.

It was already late evening by the time we arrived, and we were losing daylight.

Figure 171 group photo with Pak Zaini on the far right.

Pak Zaini (see photo far left) is a beginner breeder. Starting from his personal experience, about three years ago, he suffered from gastric disease and entered the hospital within one year, but his condition did not improve. Then he had the

Figure 170 Zaini's cow shed: Group photo with the local community leaders.

Figure 169 Hive boxes on racks at his cow shed. The Bees in east Jave seem quite compatible with cows and oxen.

initiative to consume klanceng honey taken from a bamboo colony nesting on his roof. After consuming Klanceng honey, he felt his condition was getting better, which prompted him to raise Klanceng bees. Pak Zaini took the colony in the bridge used for the roof of his cow-pen. Then, it was moved to the box from the bamboo colony. He has ten colonies (5 still in bamboo, and five have been moved into box hives). Some obstacles have also been experienced in raising livestock, such as lack of vegetation so that honey is very minimal. Some colonies also do not develop because of ant invasions, and temperatures are too hot. Although the honey yield is minimal, Mr. Zaini still sells honey obtained from his fellow breeders at 115rb-120rb per 250 ml size bottle. In addition, he also sells yellow *Xanthostemon* flower seeds and *Batavia* flowers which farmers in Klanceng often seek for 10-15 thousand per seed. His house's vegetation is yellow *Xanthostemon* flowers, *Batavia* & *Jatropha* flowers, banana plants, star fruit, etc.

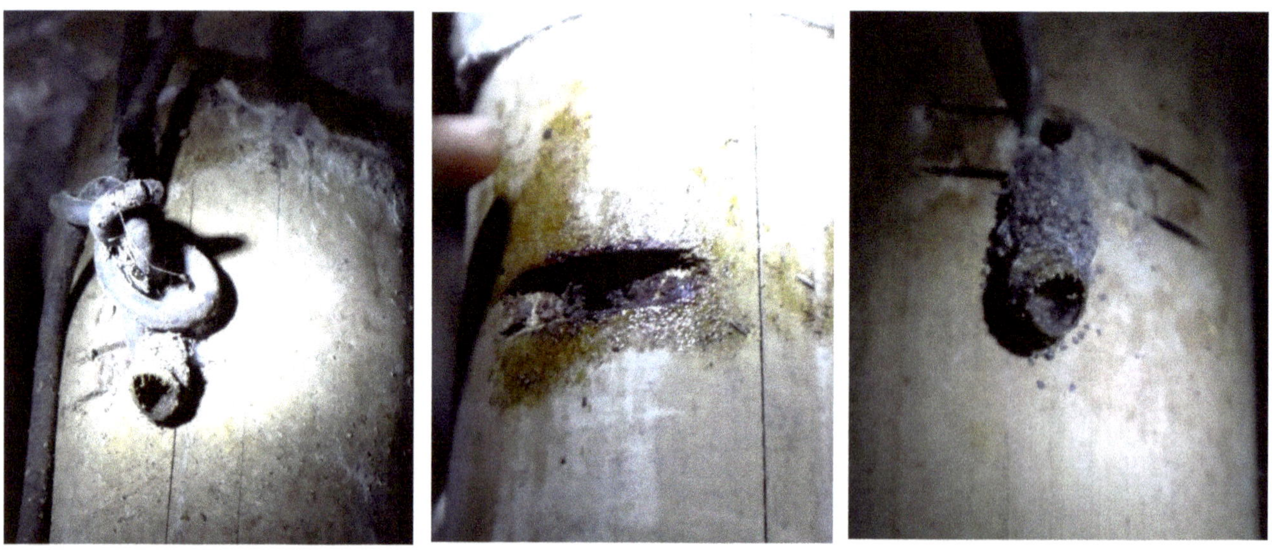

Figure 172 Tetragonula spp hives in the Bamboo pillars of the shed.

The name Tulungagung comes from two words, tulung and agung. Tulung means great source, while Agung means big. In Javanese language interpretation, Tulungagung is an area that has a large water source. Before the Japanese occupation, the Niyama Dam in South Tulungagung had the only water sources anywhere in the Tulungagung area.

Previously the capital city of Tulungagung was located in the Kalangbret area and was given the name Kadipaten Ngrowo (Ngrowo also means water source). This naming was when the capital city of Tulungagung began to move to its present place. This movement occurred around 1901 AD.

Tulungagung is well-known as one of Indonesia's largest marble producing areas and is located 154 km southwest of Surabaya City, the capital of East Java Province. Not surprising that Pak Zaini presented me with a Marble ashtray as a souvenir.

Chapter 2-11

2-11.0 - Madura Island

We continue today onward from Kediri onto Kamal, Bangkalan, Madura island. We thank Mr. Terry Yulianto at Wonogiri for the special hospitality and Mr. Seno (even though I didn't have time to meet). Many thanks to Mr. Harry and his family in Trenggalek (not forgetting his friends, Mr. Wito Kopi N Gang). Thanks to Mr. Zaini and his friends in Tulungagaung. We arrived at Surabaya early morning and proceeded to the jetty in Tanjung Periok to catch the ferry to Madura Island

Figure 173 Ferry terminal @ Surabaya port

2-11.0.1 - Pak William Kaawoan @ Bangkalan, Madura, East Java

Figure 174 Group photo with Indra (far left) and William (far right).

Mr. William (see Photo far left) is a 3-year veteran breeder working in klanceng. He has 100 colonies of *Tetragonula laeviceps*, 5 colonies of *Lepidotrigona terminata*, 10 colonies of *Tetragonula biroi* and 15 colonies of *Heterotrigona itama*. In addition to Madura, he also has 700 *T. laeviceps* colonies in West Java and 15 *H. itama* colonies in Kalimantan. Pak Wiliam harvests bees for about 2-3 months in the dry season, but once every 4-5 months if the rainy season arrives. Of the six colonies, *T. laeviceps* can produce about 1 kilogram, with honey produced sold at 375 thousand per one kg, 200 thousand per 500 grams and 125 thousand for 300 grams.

Figure 175 Pak William's Honey Brand.

During breeding, the constraints faced were a lack of vegetation because he raised the bees in the living area. Besides that, predators such as lizards, black beetles, dragonflies, serpent birds, and ants were having negative impacts. In addition, Mr. William also sells colonies in bamboo for Rp75-100 thousand per bamboo and the price per colony in the Rp150 thousand boxes. Mr. William also actively shared tips and tricks about how to breed, split or join the colonies, harvest honey, and much more that we can see on his YouTube channel "MasBro Trigona Klanceng".

The interesting thing is how he grafted colonies from colonies into log boxes. He uses a transparent hose attached to the log entrance and entrance box. Then he makes another entrance in the wooden box for the entrance and exit of the bee. He claimed the success of this technique was around 95%, but the disadvantage was that it took a long time, around 3-4 months, until the colony built its "house" back in a wooden box and received a new queen.

This view is the facade of his house, where he also has a retail outlet selling his Bee derivative products and Stingless Bee Honey. His verandah cum living room also keeps some colonies on shelves.

More on this site:

https://www.youtube.com/watch?v=f0q4hDh4Dks&feature=youtu.be&fbclid=IwAR3VUjAVMWrA9SX6rpbBc3ZkCYLA4OsbssnxnZJlOWSjq8ZVBOeDSyRujms

Figure 176 Top left: Bamboo Hives; Mid Left: Rack of Box hives; Bottom left: Williams Backyard vegetation; Right column: peculiar nest entrances.

Figure 177 Pak William adjusting his eduction setup.

Photo Album from Dr. Bajaree Chuttong during Madura trip

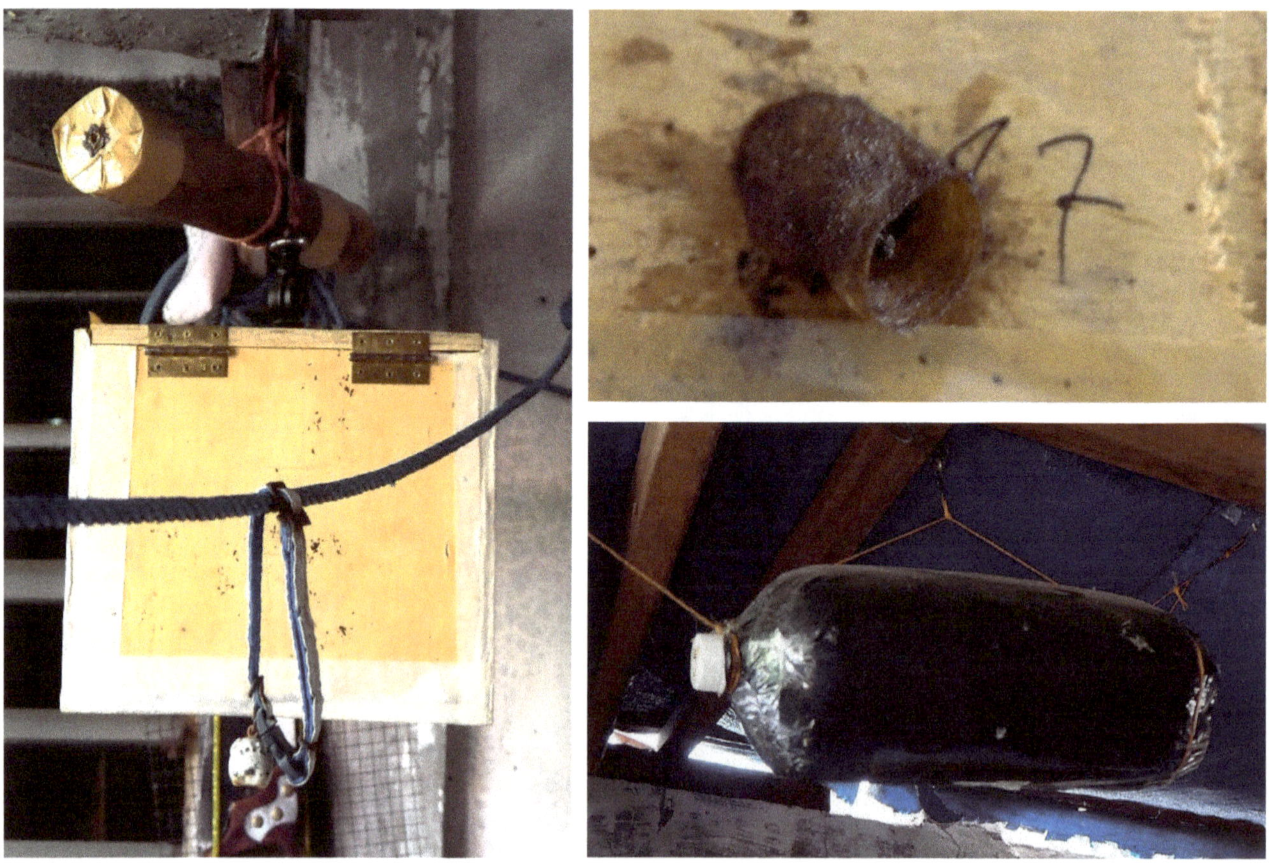

Figure 178 Experimental eduction and swarm trap.

Figure 179 Top left: Still uses plywood boxes; Top Right: Stack of Tetragonula *sp. hive boxes; Bottom left: William's Bee Products shop front; Bottom right: Shelves at his veranda living room.*

2-11.0.2 - Pak Indra @ Surabaya, East Java

Figure 180 Left: with Indra in Surabaya; Right: Lunch with Indra in Madura

Pak Indra (See Photo 2nd from right) is a *Heterotrigona itama* or klanceng beekeeper from Central Kalimantan but has 1 *H. itama* colony still in the log in Bangkalan Madura. The main objective of raising livestock is to sell *H. itama* honey. Four years ago, based on his experience, he brought the *H. itama* Colony from Sibolga and *Tetrigona apicalis*, but the colony died and was destroyed when it arrived. *Heterotrigona itama* in his house only lasted a few months in Surabaya because of the non-supporting vegetation. Then Mr. Indra chose to breed *H. itama* in Central Kalimantan, which considered the vegetation there and abundant *H. itama* colonies. Pak Indra has 30 colonies of *H. itama* in log hives. The honey is sold for 600 thousand per 500 ml. Besides that, Pak Indra also sells equipment for cultivation or the process of harvesting honey. Some of the obstacles faced were the bee colonies (*H. itama*) were attacked by predators such as ants and lizards.

Figure 181 Indra's Honey Label

His family has a core business of Real Estate in Indonesia, and Pak Indra has many business interests, especially in beekeeping accessories and Honey. Pak Indra also develops a beekeeping community in Sampit, Central Kalimantan, which we will be visiting soon. He is reasonably creative and has many unconventional ideas, although sometimes rather loud.

This is an "out of the box" idea for his Honey sticker label. He finds setting base in Surabaya has more commercial significance.

Surabaya City is the capital of East Java Province, Indonesia, and the largest metropolitan city in the province. Surabaya is the second-largest city in Indonesia after Jakarta. The city is located 796 km east of Jakarta and 415 km northwest of Denpasar, Bali. Surabaya is located on the north coast of the eastern part of Java Island and faces the Madura Strait and the Java Sea.

2-11.0.2 - Pak Opini Sumenep. Madura, East Java

Figure 182 Pak Opini's meliponary location in Sumenep

A scene at Pak Open's place in Sumenenp before the earth tremor. Although we were scheduled to visit his place, we were diverted because it so happened that the President of Indonesia was coincidentally there (in Madura) to officiate the opening of the new Suramadu Bridge that connects Surabaya with Madura Island.

We did not make it to the Sumenep because Pak Opini's place was quite badly affected, and he had to sort out and rearrange the rest of his farm after the earthquake.

This earthquake, on Oct. 11 2018, was 6.0 on the Richter scale (see attached map). The area worst hit was in Sumenep Madura (Sapudi Island), although the epicentre was in the middle of the Madura Straits.

It doesn't reflect major devastation on Sumanep because of this tremor as the epicentre was 55 km from Situbundo, East Java.

The shock was more pronounced in Sumanep ... three human lives were lost.

Figure 183 Devastation after the earthquake.

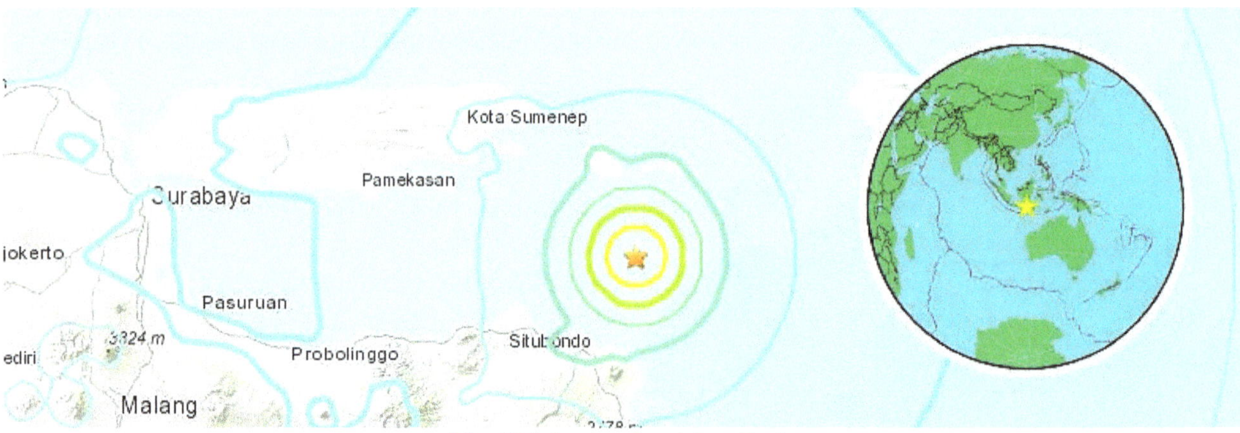

Map 10 Epicentre of east Java Quake.

Chapter 2-12

2-12.0 - East Java

2-12.0.1 - Pak Aji Said @ Desa Jebreng Lor, Probolinggo, East Java.

Pak Aji has 23 hived colonies (in the box) and inside bamboo and 2 *Heterotrigona itama* colonies in logs purchased from Lampung. Pak Aji is a Klanceng breeder who has just started in the last two months and only knows the efficacy of Klanceng honey. Previously, around one year, Pak Aji had been actively selling *Apis mellifera* honey from kapok and rubber flower nectar sources but only found klanceng honey for the

Figure 184 Top Right: Pak Aji (in pink T-shirt) in front of his residence. Below are the bamboo hives and box hives of Tetragonula sp.

past 4-5 months.

The average price of honey of *A. mellifera* is sold at around 70rb / kg and one can sell 20-25 kg per month. In comparison, klanceng honey is sold at around 150,000 rupiahs per bottle, measuring 460 ml. He claimed he could sell around 7 kg/month. The *H. itama* colony also appears to be not well developed, allegedly due to lack of vegetation and disturbing pests such as beetles, ants and lizards.

Geography: Probolinggo City is a city in East Java Province, Indonesia. Located about 100 km southeast of Surabaya, the city of Probolinggo is bordered by the Madura Strait in the north and Probolinggo Regency in the east, south and west. Probolinggo is the fourth largest city in East Java

after Surabaya, Malang and Kediri. The city is located in the Tapal Kuda area, East Java and is the main north coastline that connects Java to Bali.

Etymology: Probolinggo, which has to do with ancient stories, namely the fall of a luminous object (meteor). The fall of the object by kings was chosen to obtain peace and end disputes.

Map 11 East Java route and Location of probolinggo

Probo in Sanskrit means light, while

Figure 185 Top left: Sharing some meliponiculture tricks wih Pak Aji; Top right: Some snacks and fruits served to us with coffee.

Lingga means sign, a sign of peace. It can also be interpreted: as original or simple (like the embodiment of all simple symbols)

2-12.0.1 - Pak Nur Hidayat @ Desa Sumber Jeruk, Kec. Jambesari Darussalam.

Figure 186 Left: Pak Nur Himayat's Brand "Ijen Stingless Bees Farm"; Right: Abu Hassan with Pak Nur Hidayat

We arrived in Bondowoso district and proceeded to Pak Nur Himayat's meliponary. Pak Nur Hidayat started beekeeping on the encouragement of his friends and neighbours almost one year ago. Even so, we found his knowledge and efforts to provide the right amount of vegetation were quite encouraging. Lots of vegetation in Pak Nur Himayat's house like Durian, Mango, Sunflower, Calliandra and *Antigonon leptopus*, etc.

Mr. Nur Hidayat has for the past year been raising Klanceng honey with around 300 colonies in wooden boxes. But the distance between the colonies in the box placed on the wooden shelf is too close. In addition to raising their livestock, Pak Nur Hidayat also sells Sulawesi honey at 100 thousand per 250 ml or 400 thousand per 1 kilogram. Nur Hidayat also has 1 *Heterogrina itama* colony in the log that looks not well developed. There is no visible activity of bees coming in and out. The price of the colony purchased also varied; *T. laeviceps* was bought at a price of around 50 thousand, *H. itama* was 1 million rupiahs, and *T.* cf. *biroi* was 800 thousand rupiahs (from Sulawesi). The wood used to make boxes is from Sengon and Jabon trees. The pests were beetles (*Aethina tumida*?) and ants.

Figure 189 Group photo of visiting team with Pak Nur Hidayat and his wife

Figure 190 Although his vegetation preparations were spot on, his overcrowded placement of his hives left much to be desired.

Figure 192 Some of the flowering shrubs he provided for his stingless bees.

Figure 191 Problematic colony possibly due to overcrowding.

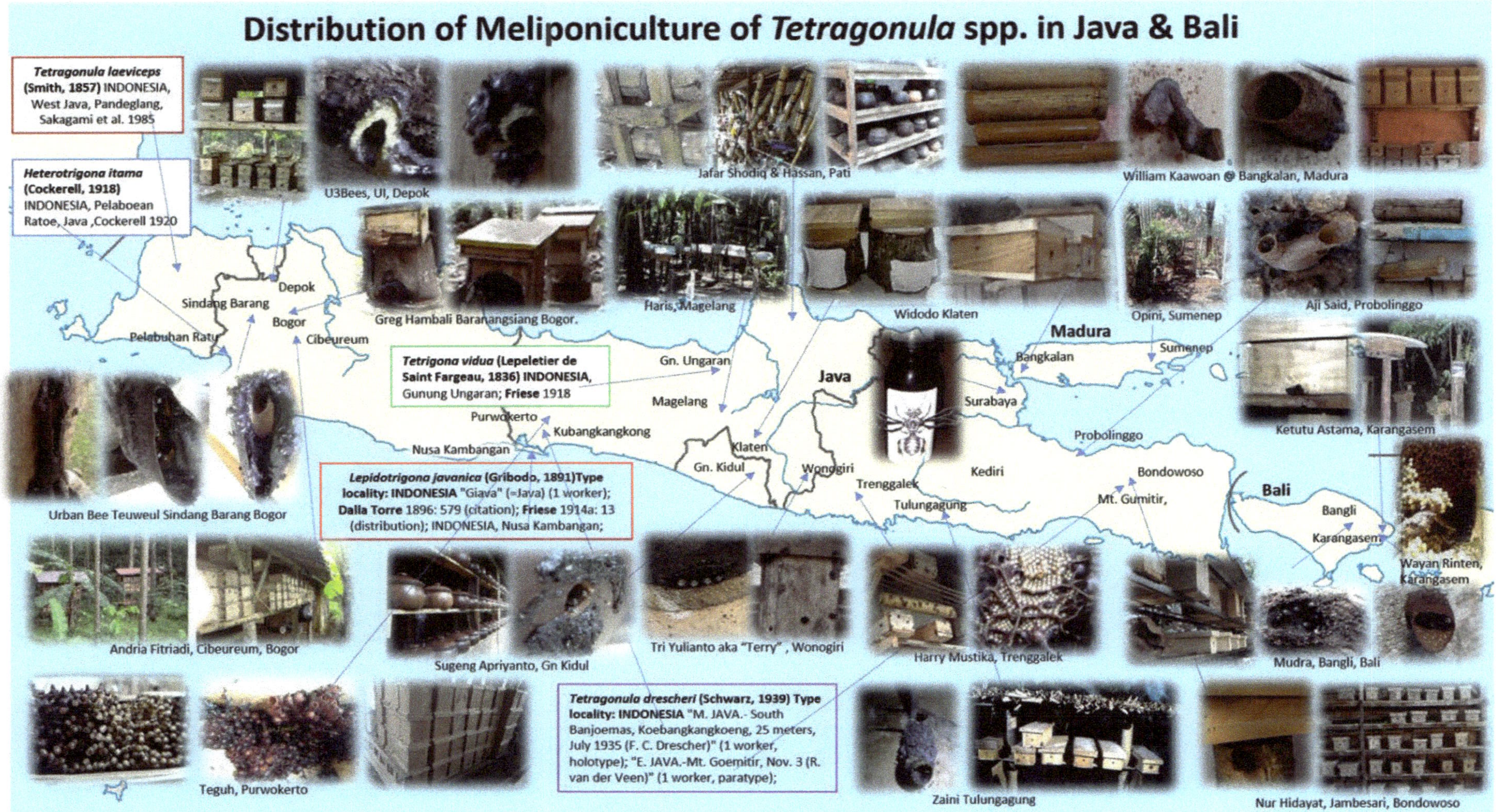

Map 12 Meliponiculture spread in Java & Bali

Chapter 2-13

2-1.0 Island of Bali –

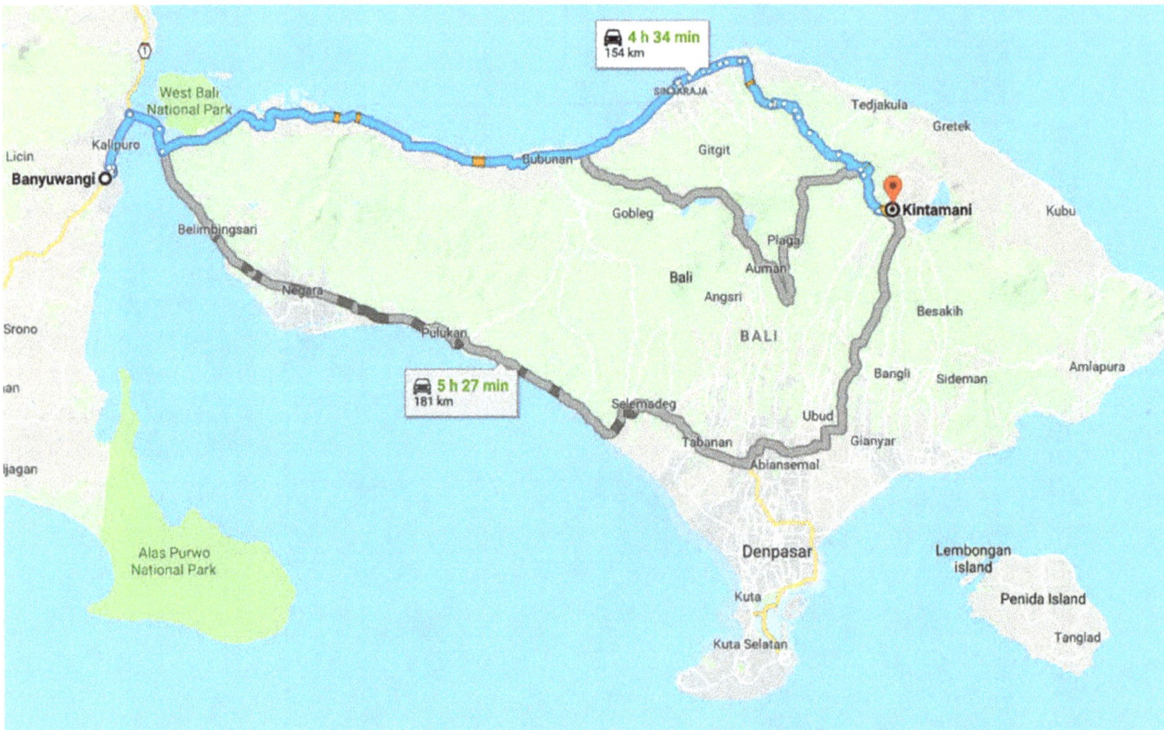

Map 13 Routes in Bali

Figure 193 The scenic view along the way from Gilimanuk to Bangli.

2-13.0.1 - Pak **Mudra** @ Jl. Putra Yudha, Lingkar Br Kawan, Kec Kawan, Kab. Bangli, Bali.

We left Bondowoso in the evening and arrived at Banyuwangi to put up the night before taking the ferry across to Bali the next morning. The crossing was about an hour, and we took the northern coastal drive of Bali. It was quite a scenic route, as seen in the photographs

We arrived at Kintamani and stopped for rest and drank in the beauty of the landscape. Continued with a rather winding road uphill.

Pak Mudra has cultivated kele bees for about four years (called klanceng in Bali). He has a colony of *Tetragonula* cf. *laeviceps* and *Tetragonula* cf. *biroi,* mostly taken from bamboo, are then transferred to wooden boxes. There are around 150-200 colonies in his house with vegetation around Pak Mudra's house, which is still very natural, allowing the Klanceng bee colonies to live and develop well. The beginning of raising a family had begun with his father. Initially, Pak Mudra's father made a trap from a former log of dried coconut trees, hoping that the *Apis* sp. bee colony that entered the coconut log turned out to be a colony of kele bees that entered the trap, and until now, the log is still used nesting kele, the age of the coconut tree log has been 47 years (1971-2018). This account raises the question,

Figure 195 Unique orchids found at his backyard

Figure 196 Current box hives at Pak Mudra's Meliponary.

how many generations of these kele-colony bee colonies are in the coconut logs? Mr. Mudra also experienced failure in raising livestock. He once raised *H. itama* by buying dozens of *H. Itama* colonies. But the colonies were not well developed, so Pak Mudra suffered losses of almost 30 million because all of his colonies were destroyed. Pak Mudra sells kele honey for 150-200 thousand per 250 ml obtained from farmers outside Bali. He hoped that the stock produced by his livestock could be sufficient for honey to be sold later.

Figure 199 Some of his vegetation and racks in his front courtyard.

Figure 203 Box hives covered with linoleum to protect from rain and hung on trees to reduce access to predators.

Figure 202 Above: Hollowed log colony; Below: PVC pipe hive.

Figure 204 Pak Mudra's Front courtyard; Middle right group photo (from left Mudra's wife, Felicia Tjoa, Bajaree, Norita, Mudra's mother, Abu Hassan, Mudra's father, Pak Mudra & Daniel)

Figure 205 Some photos taken of interest to Dr. Bajaree. Bottom left shows the typical entrance to a Balinese house.

Photo album from Bajaree

Figure 206 Variety of colonies and some even placed at his dove coop.

2-13.0.2 - Pak Ketut Astama @ Desa Juntal Kaja, Kec. Kubu, Kab. Karangasem, Bali.

Figure 207 Arrived at Pak Ketut Astama residence and farm by night and the only garden illumination was from the lights at his rest hut.

We were held back at the previous location unexpectedly, so much so by the time we arrived at this location by nightfall. We could not survey much of Pak Ketut's farm but gathered some images with flash.

Pak Ketut (see photo far right) is an *Apis cerana* beekeeper who rears oxen. Living on an area of 1 ha with unspoiled vegetation, so that colonies of 80 kele and 40 of *A. cerana* lived and developed well. Interestingly, none of the kele colonies was obtained from bamboo trees, but most were just trees and

Figure 208 Pak Ketut briefing us on his farm activities.

a little from mango trees. In one village, only Pak Ketut is raising bees. His experience, initially only raising bees *A. cerana*, then learning from Google how to breed kele, made him try this because they were easier to cultivate. From the results of raising them, Mr. Ketut can sell honey from kele kele and *A. cerana*, but the people there prefer the taste of *A. cerana* honey compared to kele kele. The obstacle faced is to market the honey. Honey of 260 Rp per 1 bottle of 600 ml and the price of honey *A. cerana* 'multiflora' are sold at 200rb per 1 bottle (600 ml).

The vegetation around the house is cashew, mango, acacia, kranci and many more wildflowers. I had a taste of the stingless bee honey dominated by the Cashew Nut portion of the farm, and it was delicious. What's left is to bring home some and get some lab tests done on it.

Seems like this secluded part of Bali (although lacking street lights) has a Shangri-La of exquisite Honey and Agro-produce. Though the soil does not appear overly fertile, Pak Ketut and his father keep their crops [thriving]. well boosted.

Figure 211 Pak Ketut's Bee farm by flash light

2-13.0.3 - Pak Wayan Rinten, Karangasem, Bali.

Figure 217 Pak Wayan Rinten harvesting at request of his customers, namely local enforcement officers.

- Number of colonies = thirty
- Honey harvest = once a year
- Harvest method = squeezed & sieve
- Length of raising = five years
- Price of honey is two hundred sixty thousand rupiahs per six hundred ml.

The information and images were obtained at a later stage. Due to the late arrival at the previous location, we could not visit Pak Wayan's farm. His 30 colonies are kept in Bamboo Hives.

Figure 214 Pak Wayan Rinten with his customers.

Figure 220 Pak Wayan Rinten has 30 colonies in Bamboo hives, these are some of them.

2-13.0.4 Bibliography Section 2 (contributed by Dr. Bajaree Chuttong)

Bogdanov, S. (2009). Honey Composition. *Bee Product Science*.

Bogdanov, S., Vit, P., & K1ilchenm, V. (1996). Sugar profiles and conductivity of stingless bee honey from Venezuela. *Apidologie 27*.

Boontop, Y., Malaipan, S., & Wiwatwittaya, K. C. (2008). Diversity of Stingless Bees (Apidae: Meliponini) in Thong Pha Phum District, Kanchanaburi Province, Thailand. *Kasetsart J. (Nat. Sci.) 42*, 444 - 456.

Chanchao, C. (2009). Antimicrobial Activity By Trigona Laeviceps (Stingless Bee) Honey From Thailand. *Pak J Med Sci April - June 2009 (Part-II) Vol. 25 No. 3*, 364-369.

Chinajariyawong, A., & Saiboon, S. (n.d.). Culture of the stingless bee, Trigona laeviceps Smith (Hymenoptera: Apidae). *Institute of Agricultural Technology, Walailak University*.

Cohn, D. (2004). Xunan Kab, The Stingless Bees of the Yucatan: Preserving meliponiculture in Mayan communities. *American Bee Journal*.

Couvillon, M. J., Wenseleers, T., Imperatriz-Fonseca, V. L., Nogueira-Neto, P., & Ratnieks, &. F. (2008). Comparative study in stingless bees (Meliponini) demonstrates that nest entrance size predicts traffic and defensivity. *J. Evol. Biol. 21*, 194–201.

Danaraddi, C. S., Viraktamath, S., & Bhat, K. B. (2009). Nesting habits and nest structure of stingless bee, Trigona iridipennis Smith at Dharwad, Karnataka. *Karnataka J. Agric. Sci., 22(2):*, (310-313).

Dejtisakdi, W., & al., e. (n.d.). Melliferous plants for the stingless bees (Trigona collina Smith) in deciduous forest of Queen Sirikit Botanic Garden, Mae Rim, Chiang Mai Province. *Department of Botany, Faculty of Science, Kasetsart University*.

Eltz, T., et al. (2003). Nesting and Nest trees of stingless bees (Apidae; Meliponini) in lowland dipterocarp forest in Sabah, Malaysia with implications for forest management. *Forest Ecology and Management 172*.

Jongjitvimol, T., Boontawon, K., & Deowanish, W. W. (2005). Nest Dispersion of a Stingless Bee Species, Trigona collina Smith, 1857 (Apidae, Meliponinae) in a Mixed Deciduous Forest in Thailand. *The Natural History Journal of Chulalongkorn University 5(2): 69-71*.

Oddo, L. P., & al., e. (2008). Composition and Antioxidant Activity of Trigona carbonaria Honey from Australia. *Journal Of Medicinal Food*.

Rasmussen, C. (2008). Molecular Phylogeny Of Stingless Bees: Insights Into Divergence Times, Biogeography, And Nest Architecture Evolution (Hymenoptera: Apidae: Meliponini). The *University of Illinois at Urbana-Champaign*.

Rasmussen, C., & Camargo, J. M. (2008). A molecular phylogeny and the evolution of nest architecture and behavior in Trigona s.s. (Hymenoptera: Apidae: Meliponini). *Apidologie 39.*

Rattanawannee, A., & Chanchao, C. (n.d.). Bee Diversity in Thailand and the Applications of Bee Products. *Department of Biology, Faculty of Science, Chulalongkorn University, Bangkok, Thailand.*

Roubik, D. (2006). Stingless bee nesting biology. *Apidologie 37.*

Sawatthum, A., & Tadakittisarn, P. V. (n.d.). Comparative composition of honey from Thai stingless bee and European honeybee (Apis mellifera L.). The *Rajamangala University of Technology.*

Souza, B., & al., e. (2006). Composition Of Stingless Bee Honey: Setting Quality Standards. *INCI v.31 n.12 Caracas dic.*

Thummajitsakul, S., Klinbunga, S., Smith, D., & Sittipraneed, S. (2008). Genetic diversity and population structure of Trigona pagdeni Schwarz in Thailand. *Apidologie 39*, 446–455.

Vit, P., Bogdanov, S., & Kilchenmann, V. (1994). Composition of Venezuelan honey from stingless bees (Apidae: Meliponinae) and Apis mellifera L. *Apidologie 25.*

Vit, P., Medina, M., & Enríquez, M. E. (2004). Quality standards for medicinal uses of Meliponinae honey in Guatemala, Mexico and Venezuela. *Bee World 85(1): 2–5.*

Vit, P., Oddo, L. P., Mbarano, M. L., & Mcejias, E. S. (1998). Venezuelan stingless bee honey characterized by multivariate analysis of physicochemical properties. *Apidologie 29.*

Vorlová, L., & Âelechovská, O. (2002). Activity of Enzymes and Trace Element Content in Bee Honey. *ACTA VET. BRNO.*

Weirich, G. F., Collins, A. M., & Williams, V. P. (2002). Antioxidant enzymes in the honey bee, Apis mellifera. *Apidologie 33.*

Section 3 - Exploring Kalimantan

by Abu Hassan Jalil

Figure 222 Cover photo: Scene of the seafront at Sungai Duri Fishing Village, West Kalimantan.

Contents in Section 3

Prelude to Section 3

Map 14 Excursion Route for the whole Kalimantan.

I spent a night in Lundu and visited Joshua's farm in the morning. Interestingly he had a few *Syzygium aqueum* trees amongst his meliponine hives.

The scenery of Kg Biawak Border Town, walking across to the Kg Aruk, the Indonesian side of the border. Across the border, in Aruk, one can hire passage by one of these Transport companies. You sometimes have to wait until they get their passenger quota before proceeding.

Figure 223 Scenery of Kg Biawak Border Town. Walking across to the Kg Aruk, Indonesian side of the border. Across the border in Aruk one can hire passage by one of these Transport companies. You have to sometimes wait till they get their quota of passengers before they proceed.

Chapter 3-1

3-1.0 Northwest Kalimantan

3-1.0.1 Ahmad Subhan @ Kg. Bugis. Sambas, Northwest Kalimantan.

Figure 224 Ahmad Subhan plants long beans and since having his stingless bees forage and pollinate his long bean flowers, his yield has increased, and the beans are longer and fatter.

Arrived at Sambas and met up with Ahmad Subhan and his brothers. Gusriandi came down from Paloh and met us at the Sambas Bus Terminal. We proceeded to Kg Bugis, the site of Subhan's eldest brother. Dayak Melayik and Kedayans mostly inhabit sambas.

Figure 225 Subhan's backyard and a display of his bee products.

Figure 226 Harvest time @ Ahmad Subhan's Meliponary. Harvest time @ Meliponary Ahmad Subhan

Figure 227 from left: Ahmad Subhan, Ahmad Mujahid Ar., Gusriandi Paloh, Abu Hassan, Joshua Wong, Salim, Pak Eko Aruk, Ahmad Wildani.

After Sambas, at 1 p.m., we travelled by another cab towards Pontianak. Along the way, we stopped at Sg Duri for lunch. It is a fishing village with good seafront views to contemplate.

3-1.0.1 - Nasution Togos (Madu Bahta) @ Pontianak

Arrived in Pontianak by nightfall and stayed the night at Comfort Homestay about 10 min from the Pontianak airport. Pontianak depicts folklore about a lady who died during childbirth and came back to haunt those who caused her grief during her pregnancy.

For fear of this, the residents created a myth that the port was haunted with pontianaks or, as the locals call it, kuntilanak. The myth of how the city got its name was that (as told to me by Nasution Togos inset) a Javanese Sultan on his ship approached the delta of the city (then just a fishing village and small trading port) to capture the port. The Sultan, not to be deterred by his desire for capture, shot out a cannonball from one of his ship's canons and declared that the blast would have scared off whatever spirits and where the cannonball fell was where he would build a city. That is where Pontianak stands today, or so it seems.

Figure 229: Supper with Nasution Togos whom came all the way from Bonti the previous day just to meet

Figure 229 Nasution's Brand n label.

However, I researched how far this was true on a more realistic note. The local Dayak dialect had Ponti, meaning very tall tree, and anak means child. Hence Pontianak would mean saplings of very tall trees (in that region, very tall trees are of the family Dipterocarpaceae). However, one can only speculate that this has left the city with only saplings due to the rampant deforestation of the past. Therefore, the name sticks - saplings of tall trees "Pontianak". Believe what you like, but I am here for the interest of Meliponines! Here is the relation to meliponines of the Dipterocarp dependence species.

In a session of intense discussion with Nasution Togos, whose fb nick and Brand name of his honey are Madu Bahta, he hails from Bonti about 4 h away eastwards. He shared some of the images of his activities on his bee farm. Anyway, I did not have the time to look into the name Bonti, whether it is relevant to be the mother of Pontianak as it is a smaller town. It could be so if it were the other way around, where Bonti is the bigger town and "Bontianak" is the baby town.

3-1.0.2 – Zaka Prihatin (Penternak Kalimantan Barat) @ Kota Pontianak

Figure 230 Met with Zaka Prihatin at his home and Melipnary and later had a sumptuous brunch.

The next day I visited Zaka Prihatin together with Pak Pujiyono and Pak Nasurion Togos was also present with Pak Adi (fb: Madu Gatot Kaca)

Figure 231 later I had the opportunity to view Zaka's collection in his garden and bee farm in his front yard.

Figure 233 After examining Zaka's Tetragonula biroi *colony, Nasution presented me with a "Tanjak" traditional head gear.*

Figure 232 Seems Nasution is quite preoccupied with 'selfies'

For video references:

https://www.youtube.com/watch?v=UTyrjyNMWso

Figure 234 We opened a few hive boxes to examine the internal nest structure.

Figure 236 Found some beautiful miniature orchids which apparently the stingless bees do visit on occasion.

Figure 235 Assorted flowers foraged by Zaka's Bees in Pontianak

3-1.0.3 - Pak Adi, at Galang, West Kalimantan.

After a good spot of brunch, we headed for Galang.

Galang is about 60 km (1.5 h) north of Pontianak. The Keeper, Pak Eko, we visited had about 300 hives of *H. itama* in and around his backyard. The ones in the front of the house were not well arranged either. I feel this keeper was overwhelmed with the number of hives he has, so everything was in disarray, and the vegetation provided as foraging sources left much in demand.

Figure 237 Placement of a couple of hundred hives in the back yard. Unruly and with no proper directional placement.

The placement is quite disorganized, and he mentioned that the production had been gradually deteriorating over the past four months. He was attributing it to the dry weather. However, after showing how his placement and the lack of proper vegetation can cause a detrimental effect, he realizes that he needs to add more greenery. As can be seen, the front of his house (although under renovation) is quite barren.

Figure 238 Group photo in front of Pak Adi's house. from left Pujiono, Togos, Adi, Eko, Pak Abu, Zaka, Gunadi.

Chapter 3-2

3-2.0 West Kalimantan

The next day I took the early flight to Ketapang. This town was named after the abundant trees of *Terminalia catappa.* This tree is essentially a coastal species and surrounds this coastal town. Once

Figure 239 The Ketapang (Terminalia catappa)

Map 15 Sukadana in Ref to Ketapang city

the ripe fruit has been peeled, the inside has hard-shelled seeds, but broken shells can allow their consumption, like peanuts. When we were kids, we often ate the beans.

Figure 240 Top photos: Stopped at local favourite haunt for breakfast before heading for Sukadana; Bottom photos: Stopped by Desa Kuala Satong on the way, eta 20 min. Sukadana.

Figure 241 Here I am showing them the Ketapang flowers and coincidentally a tiny fuscobalteata was foraging it.

I was fetched from the airport by Agus Md. Imran and Mohd Rano brought me for some local breakfast. Later, I visited some beekeeping group in Sukadana, about 2 h away (see map). Stopped by Desa Kuala Satong on the way, eta 20 min. Sukadana.

3-2.1 - Pak Suardi aka Pak Sidi @ Desa Baru Benua Kayong, Ketapang

Here I am showing them the Ketapang flowers, and coincidentally a tiny *fuscobalteata* was foraging it. The indigenous people refer to Stingless bees, specifically *Geniotrigona thoracica*, as kengkapit in this region.

We visited Pak Suardi, a.k.a. Pak Sidi (in Blue) and Pak Rahmadi wearing a cap.

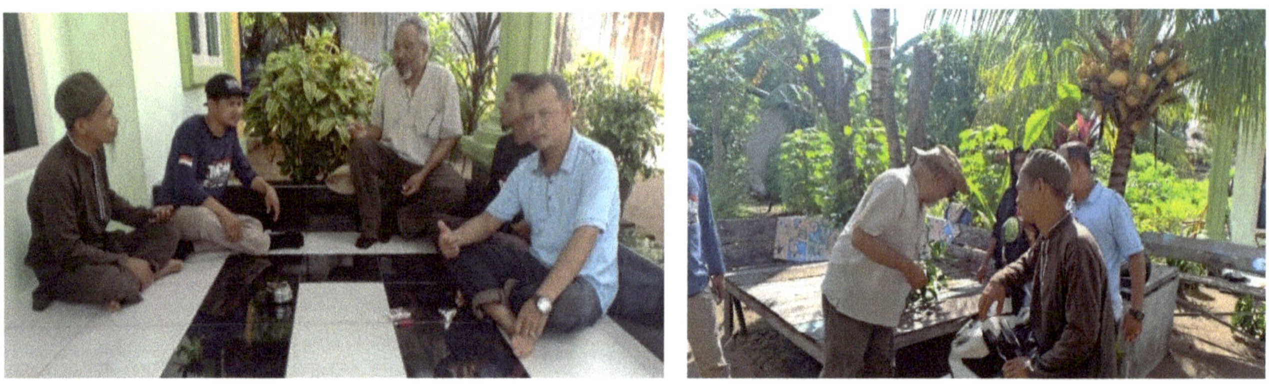

Figure 242 We visited Pak Suardi aka Pak Sidi (in Blue) and Pak Rahmadi wearing a cap. This at their home and meliponary in Desa Baru Benua Kayong, Ketapang.

3-2.2 - Agus Mohd Imran, at Dusun Tg Belimbing, Sukadana kab. Kayong utara

After this, we went on to visit some more beekeepers in Sukadana. Among them Pak Mohd Rano, Pak Juli, Pak Agoi and Pak Feriyanto all in Sukadana kab. Kayong Utara We examined Agus's eduction trials.

3-2.2.1 - History of the achievement of knowledge in Meliponiculture in Kayong utara, Sukadana district. By Agus Mohd Imran

Figure 243 One of the locations for the development of kelulut cultivation in Sukadana; Left: Honey harvesting

Figure 244 One of the places where honey farm kelulut sukadana.

The beginning of the formation of kelulut cultivation in the Sukadana district. Kayong utara started in 2017. This situation was due to the lack of public trust in the circulation of honey already sold in the field. The fake honey in West Kalimantan had first to be eliminated; in general, the cultivation of kelulut has been going on since 2005. Due to the lack of information and knowledge about the cultivation of kelulut, people are reluctant to go further in breeding kelulut. It used to be in the Selakau

Figure 245 Kelulut Sukadana honey farm at the location of the durian orchard,

area. Sambas Kalbar has one of its residents who keeps kelulut only for consumption. The wooden logs with the kelulut nests are split in two, closed again and then placed next to the house and opened when the harvest is scheduled.

3-2.2.2 - Initiation and source of colony procurement.

In cultivating kelulut, we learned a lot from the existing social media, and what made us more motivated was the arrival of Mr. Abu Hassan Jalil from Malaysia, where Malaysia itself is much more advanced in the cultivation of kelulut, compared to Indonesia. To obtain "seedlings" or colonies, we initially use the old method of taking seedlings in the forest by cutting or digging holes in the tree and then the seedlings are brought home to cultivate. However, over time we continue to learn and learn in developing kelulut colonies in an

Figure 247 *Bengkirai tree (*Shorea laevis - *Pokok Balau Kumus), one of the sources of resin in one of the honey farms kelulut sukadana*

Figure 246 Local training activities for kelulut cultivation in Ketapang district, 2019

environmentally friendly way, following what was conveyed by Mr. Abu Hassan Jalil. However, especially in West Kalimantan, only about 5% of farmers have developed many environmentally friendly colonies.

Figure 251 Broodcell ,,, leavicep. One of the types of kelulut that exists in the honey farm is kelulut sukadana

Figure 250 Heterotrigona itama layered brood cells above and nest structure with involucrum and pillars exposed

Figure 248 Tetragonula peculiar brood cell arrangement at hive box sidewalls

Figure 249 Local training activities for kelulut cultivation in the district. Ketapang kalbar,, 2020

3-2.2.3 Plant sources for kelulut feed

As for the source of plants for special kelulut feed in the west, according to our observation, the whole forest is the main source. Such as acacia trees that grow thickly, 'cengkodok/senduduk', simpur, and

Figure 252 Top Right: Zinnia elegans; *Left & bottom: Semi aquatic blossoms of Melati air (*Echinodorus palifolius*)*

others. As for the supporting plants, we provide hardy plants and soft plants. The hardy plants we planted were fruits such as mango, kweni, rambutan, coconut, petai, crystal guava, jackfruit, cassava, starfruit and other trees. We plant fruit and vegetable plants and flowers for the soft plants, such as corn, cucumbers, long beans, amp, sunflowers, spinach, kenikir and others.

3-2.2.4 Beauty and resistance in the cultivation process

Cultivating kelulut that we are struggling with today is inseparable from his situation. There is a sense of satisfaction in the breeding, and there is also a sense of frustration, all not to be missed.

There are some obstacles that we face in developing kelulut cultivation. The location of the colonies is too far away, so it isn't easy to monitor, and the rainfall is quite high. Among the problems, people still spray pesticides around the cultivation area. And the beauty is in the cultivation of kelulut when there is a lot of honey. Then the colony becomes super and a lot of demand arises for kelulut honey so that it makes an economic profit.

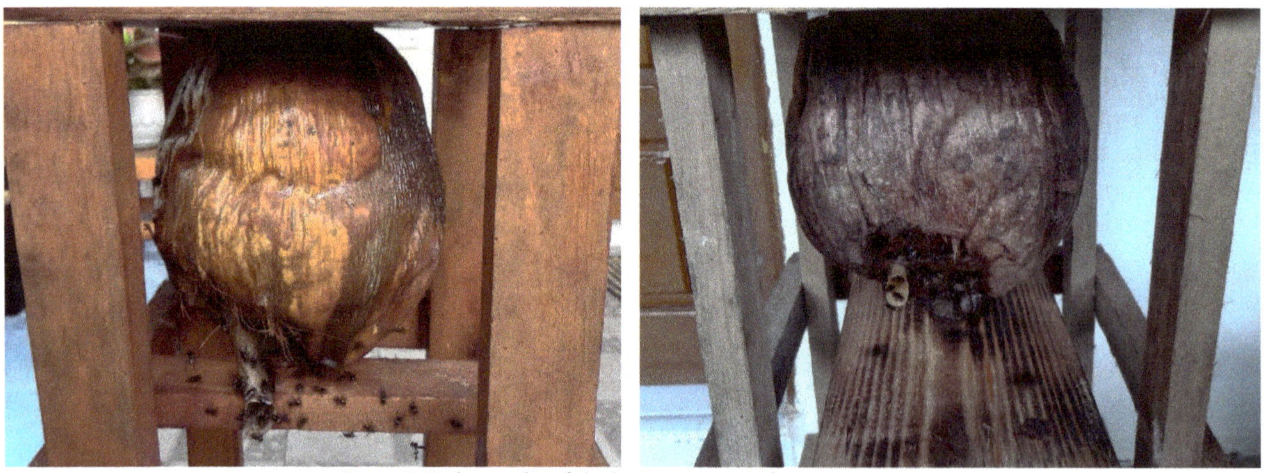

Figure 253 The results of the bait traps made for type H. Itama

Figure 254 Geniotrigona thoracica *or the local name of "Sumbu Badak" rhinoceros' axis kelulut*

Figure 255 Assorted flowers commonly foraged in Sukadana district.

Figure 256 We arrived at Pak Agus Mohd Imran's place in Sukadana.

Figure 257 Some eduction trials in Pak Agus's backyard.

Figure 258 Top Left: from left. Pak Agus, Pak Juli, Pak Abu, Pak Agoi and Pak Feriyanto all in Sukadana kab. Kayong utara; Other photos: Night discussion at a caf,e downtown Ketapang,

Figure 259 Those in attendance were (from left: Pak Agus Md. Imran, Pak Yudha, Pak Abu Hassan, Pak Tomy Yulius, Pak Dian Fernando, Pak Mohd Rano, Pak Sabirin & Pak Danie inc Pak Suadi)

Those in attendance were Pak Agus Md. Imran, Pak Yudha, Pak AHJ, Pak Tomy Yulius, Pak Dian Fernando, Pak Mohd Rano, Pak Shobirin & Pak Danie inc Pak Suadi)

3-2.3 Tomy Yulius @ Ketapang City

That night we went back to Ketapang city and had a Supper session with Tomy Yulius and the group of Ketapang Beekeepers.

Ketapang, 5/12/18

Local Blog Contributor Tomy Yulius had this to say:

Located in one of the hotels in Ketapang, the Ketapang Bee / Kelulut Bee Farmer Community held short training material on good and true cultivation. The activity also coincided with a short visit of researchers from Malaysia who had experience worldwide to conduct research and provide education to farmers.

Bpk. ABU HASAN JALIL was the name. He was very inspiring to the farmers in Ketapang and its surroundings. As expressed by Brother Agus, one of the farmers who had been quite successful and even made the Kelulut livestock business as an additional effort turned out to have produced very promising results. "The prospects are good. The future targets will try to add and multiply nests or LOGs from existing farming results", he said excitedly.

A similar thing was also sparked by one of the farmers in Tumbang Titi District, familiar with Bang UCOK, who conveyed his enthusiasm. "Indeed, this is a breakthrough that is very rare and just beginning to be enjoyed by some people, cultivators, and rarely done by the public in general, although our natural potential is still the widely available wild colonies of Kelulul Bees that exist. We haven't yet comprehended the full gist or even just the surface, even though Kelulut Honey has properties for various treatments rather than ordinary honey", he said excitedly.

In terms of discussion, a short lecture with Bpk. Abu Hasan Jalil, our area has been left behind a long while. In neighbouring countries, the cultivation of Kelulut has become an industry and many processed products derived from honey, are processed into consumables suitable for people suffering from diabetes mellitus as well as processed ingredients for cosmetics. There is no mistake the price is relatively more expensive than ordinary honey because of its proven quality.

The short lecture material is closed by hanging out at one of the "Wakops" in the city of my beloved Ales in the evening. Hopefully, more and more people will be interested in cultivating the trigona bee that produces honey from this mouth. (This blog originally in Bahasa Indonesia was posted in JW KAYONG SOLIDARITAS https://www.facebook.com/groups/1661785630566124/)

Later the next morning, I was sent off at the airport by Pak Agus & Pak Rano.

Figure 260 Left: Pak Agus sending me off at the Ketapang Airport. Right: with Pak Sidi's brother who seems mesmerized with my books. Pak Mohd Rano in the back.

Chapter 3-3

3-3.0 Central South Kalimantan

3-3.1 Laila Husaini @ Pangkalan Bun

For the following leg, I venture into Central Kalimantan territory and head for Pangkalan Bun

Pangkalan Bun was named after a small port at the Buun River. Buun river is a tributary but seems to be of economic importance.

Anyways, I was fetched from the airport by Mdm. Laila Husaini, a retired teacher, is now intent on Meliponiculture.

Figure 261 Visited Pak Rizal's Meliponary setup ata a Palm oil estate fringe.

I then visited Ms. Laila's residence. I was served some lunch and had discussions with her son, daughter-in-law and some of her neighbours.

Figure 262 Here with Arief at the airport sending me off together with Ibu Laila. Bottom Right: A view of her Meliponary showing the clean well-kept garden but lacking low flowering shrubs and ground covers.

The first visit we did was to Pak Rizal, who keeps stingless bees at the fringes of an Oil Palm Estate. Here we found the boxes of the hives not properly treated (dried), and what has happened was that internal walls were full of white mould (most likely due to wet lumber).

Figure 263 Here we found the boxes of the hives not properly treated (dried) and what has happened was that internal walls were full of white mould (most likely due to wet lumber).

Untuk kaki selanjutnya. Saya menjelajah ke wilayah Kalimantan Tengah dan menuju Pengkalan Bun

Figure 265 Examining the internal hive conditions and then explain the importance of using absolutely dry lumber

Later we visited Mdm. Laila's Residence where she kept most of her bees. She had longan trees in flower and provided many forage sources for the bees. Another beekeeper Pak Arief Gogo also joined

Herewith Arief at the airport is sending me off together with Mdm Laila.

Figure 264 Meliponary setup in Ibu Laila's Residence yard.

Looking at her place, I advised her to add more flowering shrubs and groundcovers, like portulacas and mimosas. All this, including lunch at Laila's residence, was achieved during my 3 h transit in Pangkalan Bun.

3-3.2 Pak Indra @ Sampit

I next hopped on a plane to Sampit, my other destination. Arrived at Sampit in the evening with not much daylight left. After checking into my hotel, we left for a beekeeper's home a couple of hours north of Sampit. When we arrived there, it was already nightfall, and it was raining. We had a sumptuous dinner while waiting for the rain to slow down. The people of Sampit (although a "tight" [tight = sempit, where the town got its name] quaint town are quite proud of their culinary specialities.) They like "Nasi Kuning", which is rice cooked in coconut milk set with a choice of savoury dishes like meat, chicken, fish or eggs.

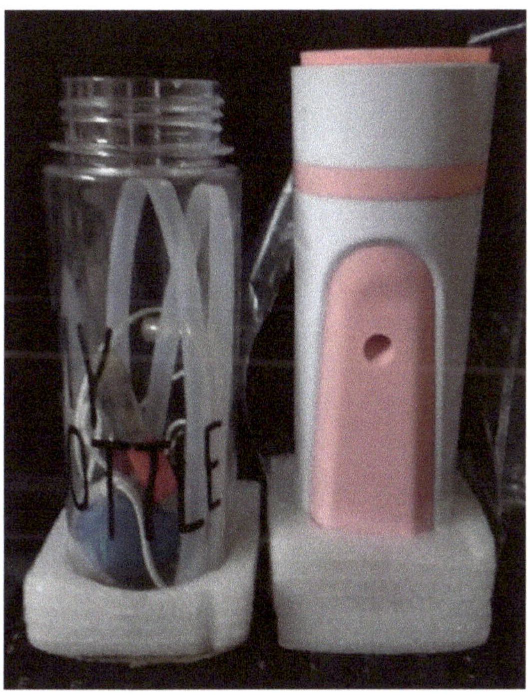

Figure 266 A suction pump courtesy of Pak Indra.

Figure 267 Indra's advert on his H. itama honey.

3-3.3 - Pak Ancah, desa Penggilingan Padi @ Sampit

It was still drizzling when we had a chance to view his hives with an umbrella and a powerful torchlight

Figure 268 Top left: with Pak Ancah's family. Bottom mid: With Indra sending me off at the Sampit Airport.

that Indra brought with him. The size of his topping boxes is rather overly large to me (about 40 x 40 x 16 cm). Mostly *H. itama* about ten hives around his wooden house. Vegetation around the house was some fruit trees with a few flowering shrubs. We were planning to visit his uncle nearby, who has some hives as well, but the relentless drizzling rain discouraged us.

Indra is the guy we met in Surabaya as he has a setup there as well, together with his family business in real estate. He sent me to the airport the next morning and presented me with a Honey suction pump. Indra then sent me back to the hotel, and we were done for the night. Overall, it was an intriguing visit as along the way back to the hotel that night, Indra stopped by a row of trees at the roadside and showed me an existing wild nest in it. Much of the Agriculture around Sampit is Oil Palm plantations.

Chapter 3-4

3-4.0 - South Kalimantan

3-4.1 - Marhadiyansyah @ Banjar Baru

My next destination was Banjarmasin, South Kalimantan, where I was met by Pak Marhadiyansyah of Banjar Baru.

Marhadiyansyah: Right for the breeders in South Kalimantan in the Banjar Baru and Marta Pura area alone. Farmers in the district join there because breeders like me in Tanah Bumbu are far away. 360 km or 7 hours drive away.

Figure 269 The mother to Pak Alan (Top left) who had just recovered from cancer (given Mr. Alan's fusco fusion honey) was very happy and could joke around.

For farmers in South Kalimantan, we have determined a route to visit so that other colleagues can follow. Tanjung Kab Amuntai Kab Barabai. We have contacted Tanah Bumbu Regency and the Kota Baru regency and are ready to gather in the new banjo, which is not yet Tanah Laut. So that he "is not tired of travelling", Kab.

We visited Pak Samlan (aka Pak Alan) in Landasan Ulin about 10 km from the BanjarBaru Airport.

The mother to Pak Alan, who had just recovered from cancer (and was given Mr. Alan's fusco fusion honey), was very happy and could joke around!

3-4.2 - Pak Fathur @ Sungai Ulin, Banjarmasin. South Kalimantan

Later we visited Pak Fathur and Pak Joko in Sg Ulin, about 15 km from Landasan Ulin. At Pak Fathur's house, his daughter paints the pink colour in Sungai Ulin, Banjarbaru, South Kalimantan.

From Left: Fathur, Bahrudin, Marhadiyansyah, Julkifli, AHJ, Iwan, Alan, Saupa. Some flowering plants in Pak Fathur's garden

Figure 271 left: from left Fathur, Bahrudin, Marhadiyansyah, Julkifli, AHJ, Iwan, Alan, Saupa. right: a view of Pak Fathur's Garden.

Figure 270 Pak Fathur's setup ata his home front garden. The pink color is painted by his daughter in Sungai Ulin, Banjarbaru, South Kalimantan.

3-4.3 - Mr. Joko place @ Sungai Ulin Banjar Baru.

Figure 272 Middle photo: Some of his trials with Styrofoam insulation sheets; Far right: Styrofoam sheet sandwiched between top cover and the box hive roof planks.

Figure 273 Some of his topping boxes are made of molded cement. Though he has not achieved the results he expects, he is still adamantly trying to improve the designs.

In the drizzling rain, we arrived at Pak Joko's place. Appears as though Pak Joko has some engineering aptitude as he comes up with assorted innovations for his Meliponine box hives.

Figure 274 Pak Joko and his friends.

Pak Joko's place at Sungai Ulin Banjar Baru. He experimented with a box made of concrete/ceramic, and there was also an experiment using Styrofoam (dotted) in the toping cover. We could not examine closer the innovative boxes because of the

Figure 275 The South Kalimantan Beekeepers gathered at Banjar Baru (the hotel where we were staying) for a discussion on the meliponiculture aspects of the area.

rain, but whatever photos we got here were taken with umbrellas. Pak Joko (see Figure 44: left-spectacled) was quite interested in discussing his innovations further to get the perspective of a foreigner point of view.

After dinner, the Beekeepers from the Banjar Baru District, Marta Pura District, Tanjung District, Amuntai District, Tanah Bumbu Regency and the Kota Baru regency gathered for spot questions and answers session. The session was led by Pak Marhadiyansyah (in a white T-shirt with a black jacket). The group was also very keen on the literature and posters I brought. We then travelled back to Banjar Baru to put up the night.

3-4.4 - Faisal Ridha @ Kec. Lokpaikat, Kab. Tapin, Prov. Kalsel

Figure 276 Faisal's family farm amongst village rubber trees. Avg output 350 ml/colony, Vegetation, rubber, banana, mango, rambutan, dragon fruit, tangerine, etc. Bottom Right: Faisal's family group photo in front of his signage.

We headed towards Rantau to visit Faisal, about 90 km from Banjar Baru. His family, parents & sisters run Faisal's farm. They have a fruit plantation that includes bananas, rambutans, longans, etc., with about a hundred bee hives.

3-4.5 - Fahrudin @ Barabai, S. Kalimantan

After rantau we headed for Barabai. The name Barabai, so the local folklore tells us, was that during the Dutch colonial days, when the Dutch were moving upriver, the locals made it difficult for the colonials by making the river impassable. They threw in logs, branches and lots of twigs, termed beraba. The act of making it happen was said as berabai and hence the name Barabai.

We arrived at the Barabai Beekeeper's association collective farm led by Pak Fahruddin.

Figure 277 The Barabai Beekeepers association collective farm.

Figure 279 Fahrudin's collective bee farm; far right: A termites nest cohabited by stingless bees.

Figure 278 Some of the nest entrances.

Figure 280 Top left: from left, Bahrudin, Faisal, Saipulah, Iiwan, Ibnu kasir, M. Yusuf, Abu Hassan, Fahrudin, M. yusri, Walat. Other photos show the Meliponary in Barabai.

3-4.6 - Fauzi (Ozie) @ Hulu Sungai Tengah, Barabai

Figure 282 fauzi's meliponary is supported by the HHBK (Non-Timber Forest Products) of the Hulu Sungai Tengah Forestry Dept.

Figure 281 A couple of Meliponine colonies elegantly presented.

we later visited Fauzi a short distance away.

Figure 283 Fauzi's Meliponary

Figure 284 Especially ornate roofing developed by Fauzy.

Figure 285 Had a brief gathering with Fahrudin's group of Beekeepers in Barabai.

That night we had a gathering and a short meet on the Meliponiculture practices in the region. They all gathered at the Hotel where I was staying.

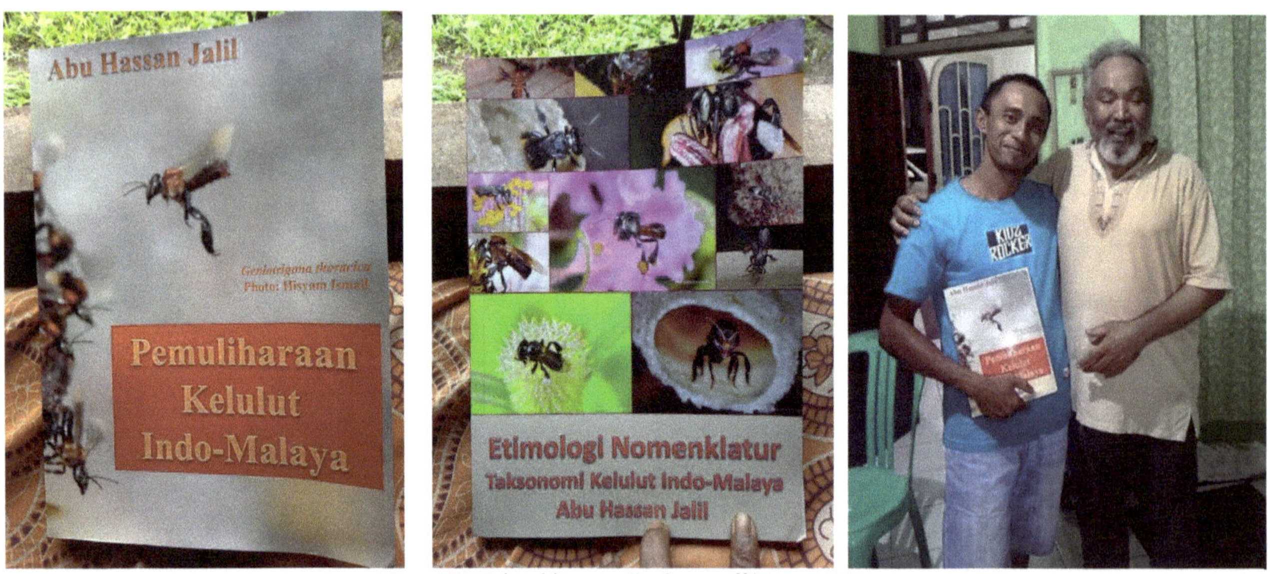

Figure 286 Fahruddin proudly shows off his acquisition.

3-4.7 - Pak Budi @ Taman Ima, Barabai

We visited another Beekeeper the next morning, at Taman Ima Barabai.

Figure 287 Pak Budi sets up a Plants nursery business in front of his house and also accommodates his meliponary in the same location.

Pak Budi here runs a plant nursery and, as a side business, practices meliponiculture. His nursery is right in front of his residence. He also makes and sells his compost and fertilizers.

Figure 288 Pak Budi's lush growth evident of his self-concocted fertilizers and compost recipes.

Figure 290 Giving Pak Budi some advice on the concepts of beescape.

Figure 289 The frontal view of Pak Budi's Nursery and meliponary; Middle photos: Lepdotrigona terminata *entrance.*

3-4.8 - Ibnu Kasir @ Binuang, South Kalimantan

Our next destination, Binuang, is a stopover on the way back to Banjar Baru Airport.

Figure 291 Pak Ibnu Kasir posing in front of his home front yard and groups' collective bee farm.

From here, we proceeded to Binaung to visit Pak Ibnu Kasir and his group of beekeepers. His group has built a dehydrating contraption that seems very simple and effective, although honey drying takes five days.

Figure 292 Pak Ibnu's Dehydrating contraption

Figure 293 Visiting Ibnu Kasir collective farm, workshop and meliponary.

Chapter 3-5

3-5.0 - East Kalimantan

Took the last flight out of Banjar Baru headed for Balikpapan. At this writing, the Samarinda airport was not functioning yet. One had to fly to Balikpapan and then take a bus to Samarinda. Spent the night for a much-needed rest and travelled in the morning to Samarinda.

Downhill brings me to Lojanan. I passed Bukit Suharto at a high altitude along the way, and the roads cut through a Dipterocarp Forest (familiar-looking trees of Resak and Keruing). The peak is about halfway between Balikpapan and Samarinda. The driver said the road was frequently blocked with fallen trees right across the road.

Figure 294 Bukit Suharto Dipterocarp Forest.

In Samarinda, I met with Dr Syafrizal Fachmi, and he fetched Prof. Siti Salmah & Prof. Idrus, and we headed for a visit to Pak Rendri, whose location is about 20 min away.

.

3-5.1 - Rendri @ Samarinda, East Kalimantan.

We were brought to Pak Rendri's farm, which had a diverse selection of stingless bee species.

Figure 295 Left: from left: Pro.f Idrus, Prof. Siti, Abu Hassan, Pak Rendri's father; Right: from left: Pak Rendri, Samarinda Beekeeper, Pak Agus Suhartanto, Pak Hartono, fellow beekeeper, Dr. Syafrizal Fachmi (with cap) & Prof. Idrus.

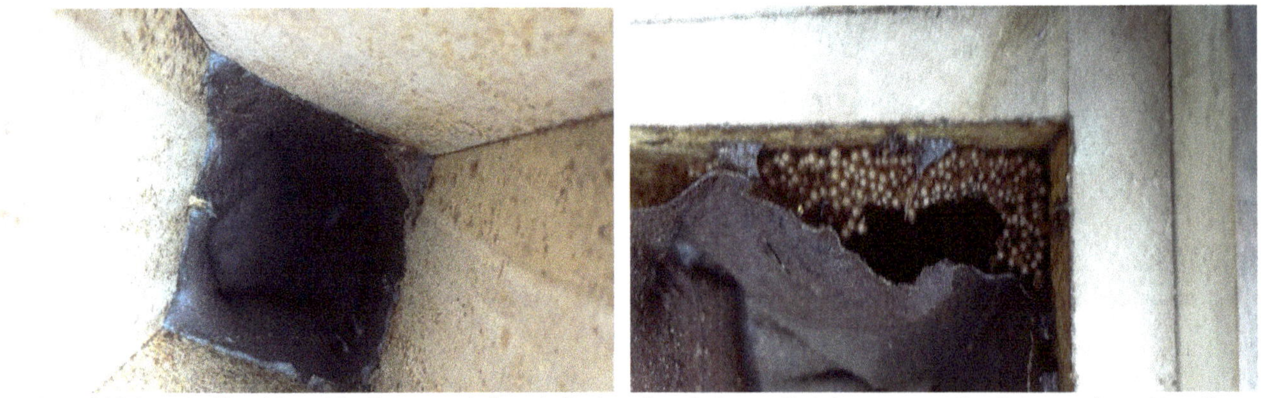

Figure 296 A vertical box hive with a Tetragonula fuscobalteata *colony just recently taken over another* Tetragonula *sp. Here the involucrum was found to be semi-hard and brittle.*

Figure 297 A collection of diverse types of stingless bees with a varied assortment of nest entrances. A. Tetragonula *cf.* minor*; B.* T. *cf* pagdeni*; C.* T. pagdeniformis*; D.* T. testaceitarsis*; E.* Homotrigona anamitica*; F.* Heterotrigona itama*; G.* Geniotrigona lacteifasciata*; H.* H. anamit*ica back vent; I.* H. anamitica *drainage vent; J.* T. sarawak*ensis; K.* T. laeviceps*; L.* T. reepeni

Figure 298 More peculiar nest entrances signifying a vast diversity of stingless bees in this area. A: Tetragonula *cf.* laeviceps*; B.* Tetragonula *cf.* giessleri*; C* T. laeviceps*;* Tetragonula cf. drescheri*;* T. *cf.* laeviceps*;* T. cf. biroi; T. *cf* reepeni*;* T. *cf.* biroi.

Figure 299 Resin type nest structures.

Figure 301 Internal structure of clustered and layered brood cells.

We were brought to Pak Rendri's *Tetragonula fuscobalteata* station of about 100 hives. The main pests

Figure 300 Rendri's Tetragonula fuscobalteata station of about 100 colonies.

he encounters are Plantain squirrels (*Callosciurus notatus*)

Figure 302 Top left: Pak Agus holding a box infiltrated by squirrels. Top right: another box devastated by squirrels; Bottom left: Group photo with Rendri's Beekeeper group; Bottom right: a cup of T. fuscobalteata *harvest from a single colony.*

Figure 303 Top photos: closeup of bees foraging Passiflora *flowers; Bottom photos: unique orchid type flowers;*

In Pak Rendri's place, I noticed mostly Calliandra and some Passiflora. The diverse species I suspect he has been depicted in figures 72 & 73.

A. *Figure 72 Tetragonula* cf. *minor; B. T.* cf. *pagdeni; C. T. pagdeniformis; D. T. testaceitarsis; E. Homotrigona anamitica; F. Heterotrigona itama; G. Geniotrigona lactiefasciata; H. H. anamitica back vent; I. H. anamitica drainage vent; J. T. sarawakensis; K. T. laeviceps; L. T. reepeni.*

B. *Figure 73: A: Tetragonula cf laeviceps; B. Tetragonula* cf. *giessleri; C T. laeviceps; Tetragonula* cf. *drescheri; T.* cf. *laeviceps; T.* cf. *biroi; T.* cf. *reepeni; T.* cf. *biroi*

3-5.2 - Pak Sugeng & Pak Purnama. Bukit Biru, Tenggarong.

Figure 304 Undulating land ideal for farming and having Meliponines for pollination

I was brought to Tenggarong to a Syzygium *sp.* (Jambu Kristal) Farm by Pak Hartono & Pak Agus Suhartanto to visit Beekeepers there. Pak Sugeng partners with Pak Purnama (Blue T-shirt).

Figure 305 Left: experimental hive hung up to avoid predators Right: Syzygium *sp. (Jambu Kristal).*

3-5.3 - Agus Suhartanto, at Tenggarong, East Kalimantan

We then visited Pak Agus at his home and farm in Tenggarong.

Figure 306 Top photos: Had the opportunity to pet a 1.5 yr old Owl; Other photos show Agus's box hives using Ulin wood,

Tenggarong is about 1 h away from Samarinda.

Figure 307 Left: with tamed owl; Right: With Agus Suhartanto and friends.

3-5.4 - Dr. Syafrizal Fachmi @ Samarinda city, E. Kalimantan

Dr. Syafrizal of Universitas Mulawarman was my host throughout my visit to Samarinda. He maintains a collection of specimens of stingless bees that are Dipterocarp dependent. He also keeps a photo gallery of some Meliponine nest entrances found at the Lempake Education Forest, East Kalimantan.

His paper is "Biodiversity and Habitat of *Trigona* at Secondary Tropical Rain Forest of Lempake Education Forest, Samarinda, East Kalimantan", 2013. The abstract is as follows:

Biodiversity and habitats of *Trigona* (*Trigona* spp., stingless bee), known as Kelulut, were studied at Lempake Education Forests of Mulawarman University (300 ha), Samarinda, which is a kind of lowland tropical rain forest. We found nine different types of Meliponines with morphological variations (*Tetrigona apicalis, Tetragonula drescheri, Tetragonilla fuscibasis, Tetragonula fuscobalteata, Wallacetrigona inscisa, Heterotrigona itama, Tetragonula laeviceps, Tetragonula melina,* and *Lepidotrigona terminata*). They have relatively small body sizes (4- 9 mm) and a pattern of black, brown, grey, orange, white and black-tipped wings. The diversity index (H'), richness index (Ri), and dominance index (C) of the *Trigona* are relatively low, which are 1.97, 1.09, and 0.16, respectively. However, the evenness species index (E') is high (0.89). The nest was marked with a building that serves as the entrance, nesting in cavities hardwood, like Ulin tree (*Eusidiroxylon zwageri*), at an open area with a maximum height of 5.65 m from the soil surface. Environmental characteristics of the *Trigona* habitat has a temperature range of 28-36 °C, a humidity of 77-96 %, a light intensity of 660-Watt m-2 (2.0-41 lux meters) and an elevation between 90-210 m above sea level.

Thanks are due to him for guiding me to the relevant sites practising meliponiculture in East Kalimantan. He also took me around Samarinda city by night lights for the last night of my excursion.

Following is an announcement from Dr. Syafrizal Fachmi on the "Tree Donor National Movement for Seeds of Rare Plants" They are free of charge, including delivery within the East Kalimantan region. GNDP will provide Free Crop Seeds to Tree Donor Friends to create a more sustainable future for Indonesia's environment.

Types of trees available:

1. *Tabebuia: Tabebuia chrysotricha*

2. Srikaya: Annona squamosa

3. Sirsak: Annona muricata

4. Bidara: Ziziphus mauritania

5. Pohon Kupu Kupu: Phanera purpurea

6. Sengon Buto: Enterolobium cyclocarpmum

7. Beringin/: Ficus benjamina

8. Flamboyan: Delonix regia

9. Kepuh: Sterculia foetida

10. Bungur: Lagerstroemia indica

11. Sawo: Manilkara zapota

12. Mahoni: Swietenia macrophylla

13. Trembesi: Albizia saman

14. Kantil: Magnolia alba

15. Sengon: Albizia chinensis

16. Puspa: Schima wallichii

17. Jabon: Neolamarckia cadamba

18. Ketapang kencana: Terminalia mantaly

19. Kedawung: Parkia roxburghii

ABOUT 30% ARE FROM THE NEOTROPICS

3-6 – References for Section 3

Ascher, J., & Pickering, J. (28 Feb 2018). *Discover Life bee species guide and world checklist*. Retrieved from Wallacetrigona: http://www.discoverlife.org/mp/20q?search=Wallacetrigona&guide=1&flags=HAS:subgenus:

Ascher, J., & Pickering. (29 Feb 2018). *Discover Life bee species guide and world checklist*. Retrieved from Tetragonula: http://www.discoverlife.org/mp/20q?search=Tetragonula&guide=1&flags=HAS:subgenus:

Eltz, T., et al. (2003). Nesting and Nest tress of stingless bees (Apidae; Meliponini) in lowland dipterocarp forest in Sabah, Malaysia with implications for forset management. *Forest Ecology and Management 172*.

Engel, M. S., & Rasmussen, C. (2017). A new subgenus of *Heterotrigona* from New Guinea (Hymenoptera: Apidae). *Journal of Melittology*.

Jalil, A. (2014). *Beescape for Meliponines.* Singapore: Partridge Publishing.

Jalil, A. (2016). *Meliponine Identifier Pictorial Guide.* Akademi Kelulut Malaysia.

Jalil, A. H. (2019). *Cohabitation of Meliponines in Ants' or Termites' Arboreal nests.* Alademi Kelulut Malaysia Sdn Bhd.

Jalil, A., & Roubik(ed), D. (2016). *Handbook of Meliponiculture.* Akademi Kelulut Malaysia.

Jalil, A., & Roubik(ed), D. (2018). *Handbook of Meliponiculture* Vol 2. *Akademi Kelulut malaysia*.

Leonhardt, S. D., et al. (2010). Stingless Bees Use Terpenes as Olfactory Cues to Find Resin Sources. *Chem. Senses 35*, 603–611.

Michener, C. D. (2007). The Bees of the World 2nd Edn. *The John Hopkins University Press*.

Moure, I. (1961). Preliminary Supra-specific classification of the old world Melliponinae Bees (Hymenoptera, Apoidea). *Studia Entomologica 4*, 181-242.

Rasmussen, C. (2008). Catalogue of the Indo-Malayan/Australasian stingless bees (Hymenoptera: Apidae: Meliponini). *Zootaxa 1935*.

Rasmussen, C., & Camargo, J. M. (2008). A molecular phylogeny and the evolution of nest architecture and behavior in trigona s.s. (Hymenoptera: Apidae: meliponini). *apidologie 39*.

Schwarz, H. F. (1937). Bornean stingless bees of the genus *Trigona*. *Bulletin of the American Museum of Natural History*.

Schwarz, H. F. (1939). The Indo-Malayan Species of *Trigona*. *Bulletin of the American Museum of Natural History*.

Syafrizal, F., Tarigan, D., & Yusuf, R. (2014). Biodiversity and Habitat of *Trigona* at Secondary Tropical Rain Forest of Lempake Education Forest, Samarinda, Kalimantan Timur. *Jurnal Teknologi Pertanian 9(1):34-38,*.

Thummajitsakul, S., Klinbunga, S., Smith, D., & Sittipraneed, S. (2008). Genetic diversity and population structure of *Trigona pagdeni* Schwarz in Thailand. *Apidologie 39*, 446–455.

Wikipedia. (17 Feb 2018). *Dipterocarpaceae*.

Section 4 - Lombok, Wallacean Region

Exploring Meliponiculture in Lombok, Indonesia

by Abu Hassan Jalil (abuhsn@gmail.com) & Norita Widya Pangestika (noritawidyapangestika@gmail.com)

Map 16 Route Map of Meliponiculture tour of Lombok Island

Contents in Section 4

Acknowledgements

Praise and gratitude, we pray to God Almighty for this Meliponiculture excerpt to be published and used for reference and a guide to beekeepers. Our main thanks go to Dr. Sih Kahono (researcher at the Indonesian Institute of Sciences) and Dr. Tri Atmowidi (lecturer at Bogor Agricultural University), who has reviewed the entire contents of this book, so that their criticism and suggestions may be helpful for this book to be completed successfully.

We also thank Mr. Nugraha Fauzan and Mrs. Nurul Wahyuni (Research and Development Center - Technology of Non-Timber Forest Products, Dr. Islamul Hadi, lecturer at Biology Department, University of Mataram) for an interesting discussion that can help us for the future to conduct further studies, and gratitude to all the stingless bee breeders, such as Pak Zulmawan, Pak Hartono, Pak Muhammad, Pak Sahlan, Pak Sukri) in North Lombok and Pak Nyoman Sutrawan, Pak Suadi, Adhe & Pak Wayan Gredeg in West Lombok for their warm hospitality.

Finally, we appreciate all the cooperation from all parties who have helped in the information and scrutiny of Lombok Meliponiculture. All constructive critiques and suggestions are also gladly appreciated for this book in approaching perfection. We hope that this book can be useful for the community.

Introduction

Intrigued by the local folklore in Lombok Island about Dewi Anjani, the Mythical Queen Goddess who oversaw Mount Rinjani's life, we learned she was mesmerised by her beauty and powers, the mystical beings of Mt. Rinjani obeyed her commands. We relate the interactions of Dewi Anjani with the pheromonic commands of the colony queen in the beehive. Such is a familiar account of what we see in the goings-on in a typical beehive.

With some reverence and much respect for this folklore, we come to Lombok Island meeting its Sasak people with their love of nature and way of living from the land. Bee honey is one of their forest resources, and this is the juncture where our interests coincide. We planned a tour to visit the village folk who use any form of meliponiculture and to study the species handled and methods.

Background

We found that Lombok is also a melting pot of different peoples of different ethnic backgrounds. We met Bima, Balinese and Javanese people besides the endemic Sasak people involved in Meliponiculture in Lombok.

At closer inspection, the sites of villagers heavily involved in meliponiculture appear to be around the fringes of the National Park of Rinjani, located at the northwest foothills of Mt. Rinjani. Some reports were found of stingless bees in the east of Lombok, but most were concentrated in the North Lombok and West Lombok regencies.

Map 17 Rinjani National Park

We visited the "Research and Development Center - Non-Wood Forest Produce Technology" in Mataram, where we met officials in charge of conservation efforts in Lombok. They currently run some programs of training villagers in meliponiculture techniques to reduce if not eradicate practices of the unruly gathering of meliponine hives from the wild.

The Indonesian Ministry of Forestry encourages matters on mitigation and adaptation of climate change through national seminars toward governance and sustainable land. Our interest in climate change affecting the bee-dependent vegetation and concentration made us look at the island's wind particle flow and velocity.

At 3 h intervals in the daytime, the wind flow seems to converge towards the northwest slopes and foothills.

(Source: https://www.windfinder.com/#10/-8.6374/116.3040/2018-05-14T06:00Z)

Hypothetically, we can assume the winds tend to converge at the northwest slopes, and so no wonder that the concentration of important flora (esp. wind-pollinated vegetation) and fauna at the Rinjani National Park is located on northwestern Mt. Rinjani.

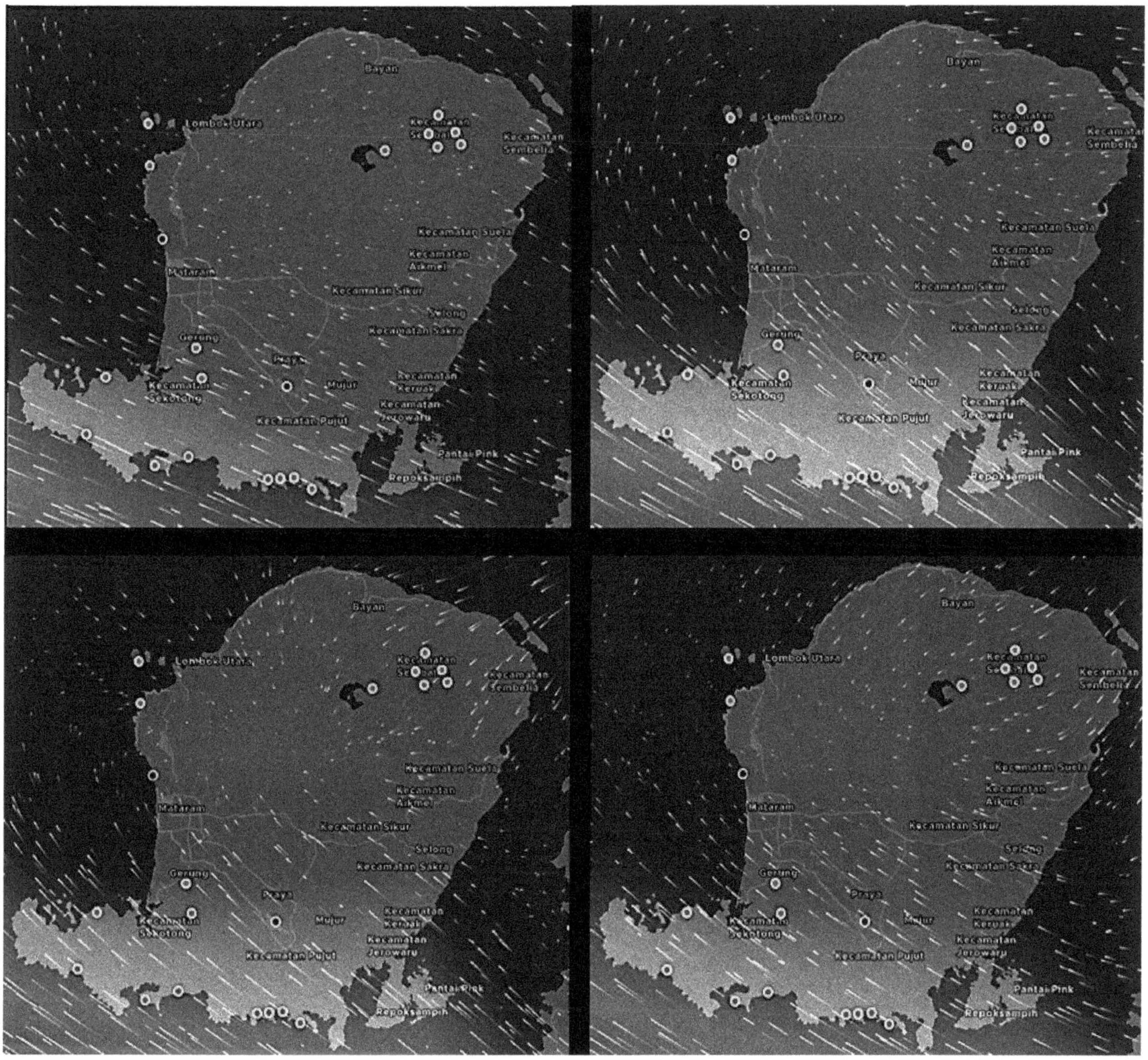

Map 18 Wind particle flow charts @ 3hr intervals on 14th May 2018.

Our interest in climate change affecting bee-dependent vegetation made us look at the island's wind particle flow and velocity. At 3 h intervals in the daytime, the wind flow seems to converge towards the northwest slopes and foothills. (Source: https://www.windfinder.com/#10/-8.6374/116.3040/2018-05-14T06:00Z)

The Windfinder information is used here under the License conditions 1.5.1 & 1.5.2 of the Windfinder Terms page (https://www.windfinder.com/contact/terms/)

We travelled by road with the tourism map as a guide to visit the Meliponiculture farms we contacted earlier via social media and chat groups of similar interest. Other maps used are https://wikitravel.org/shared/File:Lombok_Regions_map.png, licensed according to the terms in Creative Commons License, Attribution-ShareAlike 3.0.

Chapter 4-1

4-1.0 - North Lombok

4-1.1 - Bpk Muhammad Kadri & Bpk Hartono Oneman @ Gubug Lebah Group, Karang Bajo Village, Bayan District, North Lombok Regency

These gentlemen are leaders of the Meliponiculture community of about 20 households in Karang Bajo, where Hartono is the head and Muhammad is the deputy.

Figure 308 Road approaching Gubug Lebah in Karang Bajo with Mt. Rinjani in background; Main entrance to Meliponary; Tatched roof structures housing box hives amongst a forage zone of annuals and shrubs.

Gubug Lebah was formed in 2014 by Hartono, who was the head of Karangbajo village at that time. When raising Meliponines or Nyanteng, Hartono had as many as 150 colonies, with ten colonies failing. In 2016, a group of 21 members was formed. The group is open to anyone who is domiciled in the village of Karangbajo. The number of their colonies is 275, and they are cultivated in wooden boxes.

Beekeeping groups provide several types of flowers for food sources for the colonies. NOT cattle here, it is bees. Goats, sheep, cows? NO!

Plantings are of *Zinnia peruviana* & *Zinnia* spp. for pollen, *Euphorbia milli* (probably to provide resin for nest construction) & *Combretum indicum* for nectar. Palms of *Cocos nucifera* (coconut trees) abound. The main purpose of the group is to strengthen the relationship. There is the division of tasks, ease of sales or marketing, and each disseminates information on the method of bee cultivation, so the group system relies on trust. The beehive is taken from jackfruit trees, bamboo, *kesambi*, pine and timber trees. To begin a meliponary, nests existing in bamboo are collected from each household. The honey is harvested from the hive for sale. The price per 600 ml in 1 bottle is Rp150.000 with 20 litres average sales per month. In addition, the Gubug Lebah Group also sells ready-to-harvest colonies for Rp 250,000 per box. Some of the complaints were the decrease in honey production when the number of hives or colonies increased, which may be due to the feed provided as the number of colonies increased. Also, other influencing factors like the climate, fluctuating flower season, and fruiting season may be important but are not known.

Figure 309 Foraging plot of Zinnia *spp. For the bees; Discussions with UNRAM staff (Dr. Erwan flipping through books from AHJ) ; Group photo with members of the Gubug Lebah collective.*

Figure 311 Structures with 3 rows stacked with box hives; Bottom left Tetragonula cf clypearis; Bottom right Tetragonula cf sapiens.

Figure 310 Plantings of Zinnia spp. For pollen, Euphorbia milli (probably to provide sap to nest structure) & Combretum indicum for nectar. Palms of Cocos nucifera *(coconut trees) abound.*

4-1.2 – Bpk Sahlan @ Taman Lebah Meliponary, Karang Bajo sub-District, Bayan District.

Figure 312 Bamboo roofed wooden structures housing 20 to 30 hives amongst Turnera subulata *shrubbery.*

Figure 313 Gourd hive with quadrant cut-out; Pail of honey and pot-pollen recently harvested; week.

Pak Sahlan appears to be practising meliponiculture on his own. His farm is about a couple of km along the Karang Bajo road. At the front of his farm is a roadside stall selling honey from *Apis cerana* and stingless bees. We did not get the opportunity to meet with Pak Sahlan, but we did buy some honey from his wife, who operates the stall.

His farm is planted with *Turnera subulata* all over, amongst some shady trees, and his structures are made with timber with thatched roofs. He has about 60 box hives placed on three shelves (10 per shelf facing opposite directions). His wife mentioned that they have about 2,000 hives spread around the farm. That means there would be about 30 plus structures. We did not survey the whole area but saw that he did experiment with some gourds (quadrant cut-out as in the image above).

The wife told us that they just harvested their 2,000 hives which took them a total of 1 whole week and obtained a large pail full which I estimated to be about 15 litres. They also harvest the beebread, of which they obtained another large overflowing pail. It appears they harvested with the squeeze and sieve method whereas all the pollen pots appear to be crushed together. The setback is the two species they have are all mixed up.

They sell their honey at 80,000 Rp (wholesale) for a bottle of 600 ml. They also have honey from *Apis cerana* and *Apis dorsata* (locally known as 'Lani')

Figure 314 Rest hut or harvesting station and a sample Hoya flower head found growing wild across the roadside stall where they sell their honey products.

4-1.3 – Bpk Sukri (fb: Bang Slesh Apis) @ Kampung Lebah Meliponary, Sukadana Village, Karang Bajo sub-District, Bayan District, North Lombok Regency

Pak Sukri, better known as Bang Slesh, is a pioneer of beekeepers in North Lombok. Bang Slesh actively provides knowledge to the community about cultivating Teldan (the vernacular name of meliponines in North Lombok Regency). In addition, Bang Slesh also has a group of breeders of Teldan with a group of 13 members. Currently, they have 500 colonies cultivated inside wooden boxes. On average, per month, they can sell about 60 bottles of honey size 250 ml with the price of 100-thousand-rupiah bottles. In addition to selling honey, the group also provides training for children residents around Kampung Lebah without collecting fees.

Figure 315 Vegetation provided for Bee forage; Tatched roof bamboo structures house 20-30 wooden box hives.

According to Bang Slesh, a rotted bamboo structure that lasted 4 yrs. Discussion on this matter highlighted that the roofing and footing pads need to be improved to increase the structures' durability if the lifespan met their aspiration of at least culturing meliponines for ten years. The set-up organized by Bang Slesh had 12 participants who had 100 – 200 hives each. He organizes sales of the honey at Rp 100,000 for a half litre bottle.

They also liaise with Dr. Erwan of UNRAM (University of Mataram) on the bee's forage studies.

Bang Slesh is originally from Bima (east of Sumbawa, a neighbouring Island east of Lombok). He appears quite an expert on irrigation as he has managed to obtain water sources for their farm while other neighbouring areas are scorching in the ensuing drought season.

Figure 316 Pak Sukri showing his bottle of Honey AHJ bought for Rp 100k; Main Signage at the Meliponary entrance.

Chapter 4-2

4-2.0 - Northwest Lombok

4-2.1 - Bpk Zulmawan (fb: Raden Anker) @ Gengglang Village, Gangga District, North Lombok Regency

Figure 317 Arenga pinnata

Raden Anker's Facebook nickname is better known as Pak Zul in his village. He is a stingless bee hive hunter and a creative and innovative beekeeper. He hunts and collects wild bee colonies from the North Lombok Regency right to the East Lombok Regency, to districts like Masbagik & Pancor in the east, about 80 km away from his location.

He gathers most of his hives from the sago palm (*Arenga pinnata https://id.wikipedia.org/wiki/Enau*)

Figure 318 Arenga fruit stalk hives newly arrived: after sorting they are hung to acclimatize to vegetation.

Figure 319 Morinda citrifolia *log hive; Dried white gourd (Cucurbit) hollowed out: Newly arrived wooden boxes for hives.*

He employs nearby furniture workshops to make wooden box hives to transfer the *Arenga* palm hives. He sometimes finds Bamboo hives in nearby forests, and these too are transferred into boxes because the bamboos are not long-lasting.

We also found that Pak Zul used gourds and Dried Maja fruit (*Aegle marmelos*) to house his bee colonies. Historically the Maja fruit is the origin of the ancient kingdom of Majapahit (meaning Bitter Maja Fruit). When asked what inspired him to use dried gourds, he said the size and shape were similar to the beehive batumen shell. Three types were generally used, i.e. white gourds (*labu putih* or *labu bokah*); Rock melons or honeydew melons (*tawu*) and the Maja fruit. The gourd hives last quite some time, and all his initial gourd hives still exist. The only thing is preparing them takes too long, i.e. from drying (they must be allowed to overripen until dry while attached to the plant), then gouge out and sun-dry the inside for a few days. This tedious process causes him to look at wooden box alternatives.

Figure 320 White gourd hives & Maja fruit hives hung up at roof eaves amongst box hives.

Mr. Zulmawan began breeding Teldan or Nyanteng in 2013. The unique thing of how to farm is to move the nest from the tree or bamboo to the gourd or dried Maja fruit. Pak Zul has about 100 nests taken from the forest from the Arenga palm tree, bamboo, mengkudu (*Morinda citrifolia*) or noni fruit tree, and Kelanjuh wood. According to the nest obtained from the fruiting branches of palm trees and often found the nest of palm trees with a height of 15 meters from the ground. From the results of his bees, Mr. Zul can sell a bottle of honey of 600 ml with a price of 150,000 with monthly sales of about ten bottles. According to him, the yellow species (suspected *T. sapiens*) is likely to produce twice the honey of the black species (*T. clypearis*). In addition, Pak Zulmawan also sells colonies that have been moved to a box made of camellia wood. The price of an empty box is priced at Rp 25,000. Bee cultivation Pak Zul also several times had the opportunity to be covered by private TV stations in Animal World, Variety Indonesia and Spot Lite. This limelight is evidence of his conscientious cultivation of Nyanteng (Stingless bees).

4-2.2 - Bpk Adhe Adve @ Jengala village, Tanjung district, KLU

Adhe Adve is a young gentleman who keeps stingless bees with his uncle Nusno Kapu and his brothers Agus and Gede in Jengala, Tanjung, North Lombok Regency. Their first interest in the bees was umpteen years ago when they saw these stingless bees inhabited the bamboo pillars of their outdoor rest hut (aka "bale").

Then, 2 yrs ago, the uncle decided to look for wild colonies to experiment in breeding them. They found ten colonies and kept them in wooden boxes. Out of these initial 10, 2 failed, leaving 8. They were determined to keep going. They started buying more wild colonies this time (6 months ahead) from hive hunters, at about 20,000 Rp each. To date, they have nearly 160 hives. They would have had 200 hives if not for some failures. [I doubt they can abscond at all; they are NOT APIS]

- At first, instance when I visited their place, I found that their location was just adjacent to a paddy field. For paddy fields, a fixed regime of molluscicides spraying is done regularly. This regime is the major contributing factor. Other factors include placing about 100 over box hives on 2.5-meter-high racks in rows of zigzag manner. Practically no comfort zone for the colonies.

Over a 1.5 y period, they only managed (2 months ago) to get eight bottles of 600 ml honey each. The bottle of honey is sold at 150,000 Rp.

They have encountered much aggressive battling among colonies and observed many swarming activities. However, they have not gathered enough experience to recognize the differences between resource competition and mating swarms.

Much effort was made to plant ornamental and flowering plants, and again here, lack of experience (esp. Beescape) is reflected in the poor flowering density of their garden.

Figure 321 Clever hive designs wrapped with white plastic sheets.

Figure 322 Single row of wooden box hives in zig-zag manner under roof eaves.

Figure 323 Pak Adhe with his uncle Pak Nusno and his brothers; Irrigation Canal just beside his house that provides water for the adjacent paddy fields. AHJ taking photos in front of Pak Nusno's Home; Exquisite architectural stairway from the garden down to WC facilities and the canal.

Chapter 4-3

4-3.0 - West Lombok

4-3.1 - Bpk Nyoman Sutrawan @ Monjang Lestari Meliponary, Dusun Montang, Karang Bayan Village, Lingsar District, West Lombok Regency

Figure 324 Wooden box hives hung at roof eaves of rest hut (Balinese Bale); Box hives arranged 3 storeys along their perimeter side wall including one experimental log hive with a topping box at the end.

Nyoman Sutrawan is of Balinese descent and runs his farm with his 69 y old father and his wife. He says he experienced some failures of about 10-20% of the hives. He buys them in Bamboo hives (photos below).

The surrounding vegetation at his home is quite dense, and his bees rely on these for forage.

Mr. Nyoman Sutrawan is fairly new in terms of raising klanceng. He was inspired by Pak Suadi, who had already been farming. Beginning in 2018 is the initiation of Mr Nyoman's start in meliponiculture. The beginning of Nyanteng's (the vernacular name of meliponines in the West Lombok Regency) nests was obtained from the villagers' homes. On the first occasion, Mr. Nyoman had 20 colonies kept in his house compound, but four failed. After that, more colonies were added by taking from the Arenga palm forest and bamboo trees so that the number of colonies that now exist in his house was about 100 colonies. Breeding nyanteng is a part-time job because Mr. Nyoman also works in one of the hotels in

North Lombok. Like his father (16 colonies) and in-laws (12 colonies), some of his relatives also cultivate these bees at home. So that can be said this meliponary is a family concern and apply a profit-

Figure 326 Stingless bee colonies in Bamboo hives newly acquired; Dense vegetation surrounding their residence.

sharing system. Up to five months of breeding, Mr. Nyoman has never harvested honey from his colony, except for honey harvested from bamboo colonies transferred to the wooden box hives.

Figure 325 Pak Nyoman's Meliponine colony acquisition has ramped up to getting log nests while his vegetation provision has very much increased as seen from his images here in 2022.

Figure 328 The family ancestral shrine adjacent to their residence garden; Bottom left, from left Pak Nyoman, his father, the author, Pak Suadi, Pak Nyoman's wife and his two children. AHJ showing a chart of various bees and queens of Indo-Malaya stingless bees.

Figure 327 Pak Nyoman has since moved on to commercialization of stingless bee honey. These are his latest images in 2022.

4-3.2 Bpk Suadi (fb: Suu Adi), at Dusun Karang Bayan, Karang Bayan Village, Lingsar District, West Lombok Regency

Pak Suadi is the Chairman of the Meliponiculture Association or *Trigona* in Lombok. In 2010 Pak Suadi began to pursue the cultivation of Nyanteng, so it can be said he has long enough experience in meliponine cultivation. Now Pak Suadi and family rely on their living to cultivate bees Nyanteng. Pak Suadi focuses more on selling Nyanteng colonies that have been moved to wooden boxes. Mostly, the nest is taken from the fruit stalk of the Arenga palm tree, bamboo and various trees. In addition to buying and selling colonies, Pak Suadi also makes a drink of propolis extracts and bee pollen juice mixed with sweetened condensed milk and other ingredients. It took almost two years of research to ultimately produce the beverage products from pollen and propolis. As it does not have any preservatives, Mr. Suadi's blend cannot be marketed to the public as yet because it does not have a reasonable shelf life. The average income from the sale of his honey is rather low. What we observed was that he failed to provide sufficient forage sources. Due to this, Pak Suadi prefers to trade in colonies for his group. The price of the colonies was put on sale @ Rp 25,000, and the price of the colony in the box sold @ Rp 75,000. According to local research, above 400 m altitudes, only a blackish stingless bee (alleged *T. clypearis*) and less than 400 m altitudes were found brownish types (suspected *T. sapiens*). In addition, Pak Suadi also gives his extracted propolis to goats and chickens. It is believed propolis can increase body weight and boost immunity in livestock.

Figure 329 Co-author at main signage of the Karang Bayan Meliponary.

Figure 332 Stack of wooden boxes to receive meliponine colonies; Bamboo hives, Arenga hives & Box hives.

Figure 331 The welcome signage; The "Welcome drinks" Pak Suadi calls his Beebread juice (left) & Propolis Juice.

Figure 330 Peanut butter fruit, Bunchosia armeniaca, at Pak Suadi's Meliponary. He calls it Buah Madu and apparently a favourite forage among his stingless bees.

He says he currently has about 300 hives left after supplying up to 2,000 hives to a major political campaign for distribution to their constituents. Pak Suadi is known in his community as a sort of expert in *Trigona*. He receives Arenga hives from hive hunters in his village of Karang Bayan in West Lombok. He transfers these hives into boxes that he hires local carpenters to produce. He has also found meliponine hives in discarded wood pallets.

Figure 333 Tetragonula *cf.* clypearis *in Arenga hive &* T. *cf.* sapiens *Arenga hive.*

He is supported by the Regional Forestry Training Board and has innovated some derivative products, which he calls Propolis juice & Beebread Juice. This Propolis juice is also fed to his roosters, where

Figure 334 Ant rap footing of his Structure shelves; Discarded Wood Pallets found with stingless bee nests.

Figure 335 Pak Suadi feeding his rooster withhis Propolis juice with a syringe; Sorting new arrival of Arenga hives.

he claims this makes for robust growth. For his structures, he puts ant traps at the footings. He stacks the wooden boxes and wild hives at his shack with a few rooster coops.

4-3.3 Bpk Wayan Gredeg, in Selagalas Village, Cakranegara Municipality, West Lombok.

Figure 336 Woodern Stump hive with unique bark character; Hardened oozed red resin on a 'Fern Tree' trunk; Arenga Palm hive on a carved pumice block.

Figure 337 Totem pole with live meliponine nest; Tall Bamboo hives topped with erotic art; rare Bamboo cluster hive.

Figure 339 drift Bamboo rhizome cluster; Bamboo root with live bamboo hive attached; Oxen Skull with rotted drift wood trunk set on a fossilized stump.

Unlike most, this Nyanteng breeder, Pak Wayan Geredeg, is one of the famous artists in Kab. North Lombok has dozens of Nyanteng's nests in its Gallery. Some are in wooden boxes, in bamboo trees and tree trunks. The cultivation of bees Nyanteng is not the main purpose of Pak Wayan because as a "*living art*" artist, Pak Wayan makes 3D art from the honeycombs, or nest structures of meliponines can fetch a selling value up to hundreds of millions of rupiah. His work is themed living art, where Mr. Wayan Geredeg is very concerned and care about the environment. Pak Wayan is also well aware that the role of bees is very important for human life and the balance of ecosystems in the universe.

Figure 338Token of Illustrated Meliponine guide presented in appreciation of permissions to publish images of Wayan Grdeg's works and paintings; Tetragonula *cf.* clypearis *entrance on one of his Living Art.*

Occasionally, new colonies that arrive before the artwork combo harvest the surplus honey for his family's consumption.

Wayan Gredeg in Selagalas West Lombok is an artist of Balinese Descent. He collects driftwood and Bamboo cluster rhizomes providing exceptional visual character in sourcing material for his art. He combines them with live meliponine nests to invoke mystical visuals in 3D. He is also an avid painter of abstract art forms inspired by nest internal structures. His late father was also a renowned painter, and it was quite obvious to us the inherited skills flowed in his veins.[these are learned, mostly, not inherited]

Below are some of his artworks showing influences in the bee's nest structures composed of foraging sources and natural habitat.

.

Figure 340 An abstract painting depicting the sad state of bees.

Figure 341 An earlier painting of a village girl by Wayan Gredeg's late father; some bushes of Yellow Cosmos sulphurous for the bees in the courtyard; Drift Bamboo rhizome art; Internal nest structure of Tetragonula cf sapiens box hive.

Figure 342 Some paintings by Wayan Gredeg inspired by his native Lombok Stingless Bees.

Chapter 4-4

4-4.0 - Research and Learning Institutions - West Lombok

4-4.1 - **Research and Development Center - Technology of Non-Timber Forest Products** (Kementerian Kehutanan – Teknologi Hasil Hutan Bukan Kayu).

The Ministry of Forestry encourages climate change mitigation and adaptation to sustainable governance and land through national seminars. We visited the "Research and Development Center - Technology of Non-Timber Forest Products" in Mataram, where we met with Mr. Nugraha as the official responsible for conservation efforts in Lombok. They are currently running several village training programs in cultivation techniques and propagation of local stingless bees to make the existing bee colony population less vulnerable in the forests.

Figure 343 From left Ms. Nurul Wahyu, Pak Nugraha (Conservation officer) Research and Development Center - Technology of Non-Timber Forest Products and Co-author Miss Norita Widya Pangestika.

4-4.1.1 - Discussions @ LITTBANG Non-Timber Forestry technology

On the last day in Lombok, we met with local officials, Mr. Nugraha Fauzan, who was responsible for researching the conservation of beeswax in Lombok, so cultivation by the local people there did not affect much on the preservation of bees in the Forest, despite unregulated logging of trees. Therefore, the R & D of Non-Timber Forestry technology department is initiating methods of moving the stingless bee colonies from the tree without cutting down the tree. It is still an early effort to develop the technology. We did mention "Eduction" methods practised in Malaysia and Australia.

We visited the Faculty of Mathematics and Natural Science, Department of Biology, University of Mataram. We met with Dr. Islamul Hadi of Ecology and Biosystematics of Animals at Gedung FIMPA. Our interest in climate change that affects bee dependence and abundance makes us see the island's flow and speed of wind particles. We discussed this with him and expressed our intention to study this phenomenon (as mentioned in Intro. above) with any members or undergrads of their faculty.

Figure 344 From Left AHJ, Co-author Norita Widya Pangestika & Dr. Islamul Hadi of UNRAM.

4-4.1.2 - Discussions @ UNRAM

Dr. Islamul Hadi is a biology faculty member at Mataram University. Although his research background is primates (*Macaca fascicularis*), Pak Hadi has also researched Lani bees (*Apis dorsata*). A great concern towards conservation and the environment also makes Mr. Hadi care about 'nyanteng' in Lombok Island. The result of our discussion is to discuss the direction of the wind in Lombok Island, which tends to converge towards the northwest foothills of Mount Rinjani, and coincidentally concentrated at this area of Mount Rinjani, we found many farmers engaged in the activities of rearing stingless bees. One of the innovations that can also be done on the island of Lombok is to use pottery for housing the bee colonies. Similar to Pak Sugeng in Gunung Kidul Yogyakarta, pottery is one of the handicrafts in Lombok, precisely in the village of Banyumulek, Kediri, West Lombok. Earthenware clay pot (terracotta) is more durable, relatively affordable, and can minimize infiltration by natural enemies because of its closed form.

4-4.2 In Summary

Stingless bees have some local names (vernacular names), such as in Central Java with klanceng or lanceng, in West Java they are called teuweul and in West Nusa Tenggara, Teldan or Nyanteng. Each region has different species, habitats and unique ways of cultivation. As in Magelang and Klaten (Central Java), more use of wooden boxes for *Tetragonula* cf *laeviceps* cultivation where the original colonies were transferred from bamboo stems and some directly using tree trunks that already contain colonies (*Heterotrigona itama* brought over from Sumatra). Gunung Kidul, Yogyakarta province, uses terracotta vessels (pots made of red clay) for cultivation (*Tetragonula* cf. *laeviceps*) and bamboo stems. Pak Sugeng is a pioneer farmer in Gunungkidul Yogyakarta who has formed several clusters of bee colonies and gets support from the local Agriculture Dept. In addition, Mr. Sugeng also received an award from President Joko Widodo for his achievements. Different in Bogor West

Java, there is a direct cultivation by using the tree trunk (log hive) by adding a topping box above it and cultivating directly into wooden boxes.

The uniqueness of Teldan cultivation in North Lombok, West Nusa Tenggara uses the Maja fruit and pumpkin fruit which has been dried and removed its contents, but still many also use wooden hives for cultivation as in Java. In the Banyumulek area of Lombok, we observed that a centre of producing vessels or pottery could be utilised to cultivate Teldan and as a substitute for wooden boxes. This substitution was put forward to the beekeeping groups and their advisors. Another uniqueness of NTB is that there are two species of *Tetragonula* cf. *sapiens* and *Tetragonula* cf. *clypearis*. The island of Lombok is one of the islands located East of the Wallace line, so the species found on the island of Lombok are different from those in Java, Sumatra, Bali and Kalimantan. Farmer groups in North Lombok and West Lombok are also under the supervision of the relevant Government Departments in the Ministry of Forestry and the Biological Science faculty of the University of Mataram.

Hypothetically, we can assume the winds tend to converge at the north-west slopes showing no wonder that the concentration of important flora (esp. wind-pollinated vegetation) and fauna at the Rinjani National Park (located at the north-west of Mt. Rinjani slopes)

In addition, there is Mr. I Wayan Gredeg, an artist from Mataram who extremely cares about stingless bee conservation, so much so that it led him to showcase his artwork with 3D Living art. His work composes unique untainted meliponine colonies fusing with natural objects with the dramatic constitution. He does this with such fine meticulousness to generate public awareness of the importance of maintaining fauna and conserving nature in West Nusa Tenggara.

Chapter 4-5

4-5.0 - Earthquake of August 2018

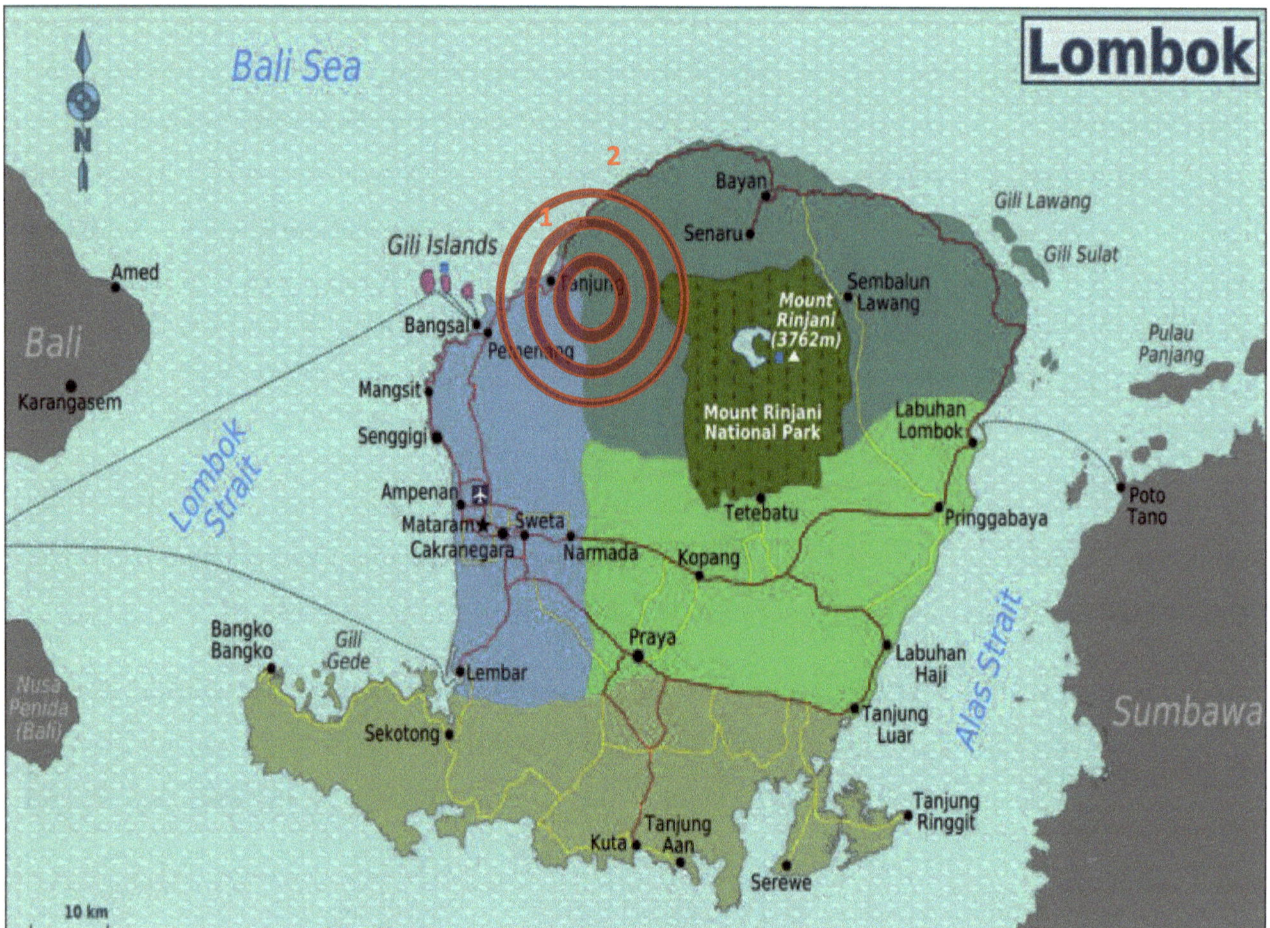

Figure 345 The latest tremor had a magnitude of 5.9 or 6.2, according to monitors. It brought down some buildings. Indonesia's chief security minister had earlier said 319 people had died, while local media have reported figures as high as 347. Source: https://www.bbc.com/news/world-asia-45124972 9th August 2018 Epicentre location reference: BMKG, BBC

The following photographs depict the devastation near the epicentre, observed in the aftermath of the Earthquake beyond 9th August 2018 at the northeast portion of the island.

1. Location: Zulmawan (fb: Raden Anker) @ Gengglang Village, Gangga District, North Lombok Regency
2. Location: Adhe Adve @ Jengala village, Tanjung District, North Lombok Regency.

Earthquakes in Lombok Island in Early August. *Gempa bumi di Pulau Lombok pada awal Agustus.*

Figure 346 After the recent Earthquakes in early August the situation has drastically changed.

Figure 347 The devastation to many villages in North Lombok.

Figure 348 This the current state as of 6th August 2018 of the home of Raden Anker.

Figure 350 This is the present situation as of 6th August 2018 for Adhe Adve in Tanjung District. North Lombok.

Figure 349 Pak Adhe's Uncle's house in shambles (See earlier photos).

4-6.0 - References for Section 4

Ascher, J., & Pickering, J. (2018, Feb 28). *Discover Life bee species guide and world checklist*. Retrieved from Wallacetrigona: http://www.discoverlife.org/mp/20q?search=Wallacetrigona&guide=1&flags=HAS:subgenus:

Baker, R. A., & Chmielewski, W. (2003). How Old Are Bees - A Look at The Fossil Record. *Journal of Apicultural Science*.

Basibuyuk, H. H., & Quicke, D. L. (1997). Hamuli in the Hymenoptera (Insecta) and their phylogenetic implications. *Journal of Natural History*.

Birt., M. (n.d.). The *Trigona* Species (Stingless Bees). Their roll in higher plant pollination. *12391833933*.

Chandra, G. (n.d.). *Wallace's Line, Weber's Line and Wallacea*. Retrieved from IASZoology.com: http://www.iaszoology.com/wallacea/

Crepet, W. L., Nixon, K., & Gandolfo, M. A. (2004). Fossil evidence and phylogeny: the age of major angiosperm clades based on mesofossil and macrofossil evidence from Cretaceous deposits. *American Journal of Botany*.

Danforth, B. N., & et al. (2006). The history of early bee diversification is based on five genes plus morphology. *PNAS*.

Darmawan, S. (n.d.). *Diversifikasi produk* Trigona *sp. di Lombok.* badan litbang kehutanan, kementrian kehutanan.

Engel, M. S., & Rasmussen, C. (2017). A new subgenus of *Heterotrigona* from New Guinea (Hymenoptera: Apidae). *Journal of Melittology*.

Goulet, H., & Huber(eds), J. T. (1993). Hymenoptera of the world: An Identification Guide to Families. *Agriculture Canada*.

Jalil, A. (2014). *Beescape for Meliponines.* Singapore: Partridge Publishing.

Jalil, A. (2016). *World Meliponine Etymology of Taxanomic Nomenclature.* Akademi Kelulut Malaysia.

Jalil, A., & Roubik(ed), D. (2016). *Handbook of Meliponiculture.* Akademi Kelulut Malaysia.

Lamerkabel, J. S., & et al. (2017). Tabiat Bersarang Dua Spesis Stingless Bee Koloni Di Pulau Ambon Muluku. *Seminar Nasional Perlebahan.* Bogor: IPB.

Raes, N., & et al. (2014). Historical distribution of Sundaland's Dipterocarp rainforests at Quaternary glacial maxima. *PNAS*.

Rasmussen, C. (2008). Catalogue of the Indo-Malayan/Australasian stingless bees (Hymenoptera: Apidae: Meliponini). *Zootaxa 1935.*

Riendriasari, S. D. (2014). Budidaya dan produk perlebahan *Trigona* spp. di Lombok, nusa tenggara barat. *Prosiding Seminar Nasional* (pp. 213 -221). Jogja: University Club Universitas Gadjah Mada.

Schwarz, H. F. (1939). The Indo Malayan Species of Trigona. *Bulletin of AMNH.*

WELZEN, P. C., J. F., & ALAHUHTA, J. (2005). Plant distribution patterns and plate tectonics in Malesia. *Biologiske Skrifter 55*, 199-217.

Welzen, P. C., Parnell, J. A., & Slik, F. (2011). Wallace's Line and plant distributions: Two or three phytogeographical areas and where to group Java? *Biological Journal of the Linnean Society.*

Xiong, C. S. (2015). A Conservation Assessment of Stingless Bees (Apidae: Meliponini) in Singapore. *Cohort AY2014/2015 S1.*

Section 5 - Sulawesi, Maluku & Papuasia

Exploring Meliponiculture in Indonesian Wallacea and Papuasia

Figure 351 Green Honey in Sulawesi

Contents in Section 5

Figure 352 Typical Tausg Architecture influence on this ancestral home.

Tausug Architectural Influence

The Tausug is a native tribe found in Jolo at the Sulu archipelago of the Philippines. •The name Tausug is derived from two words: “tau” meaning man and “sug “ or “suluk” meaning current.

Figure 353 The Spread of Tausug Architecture in Wallacea.

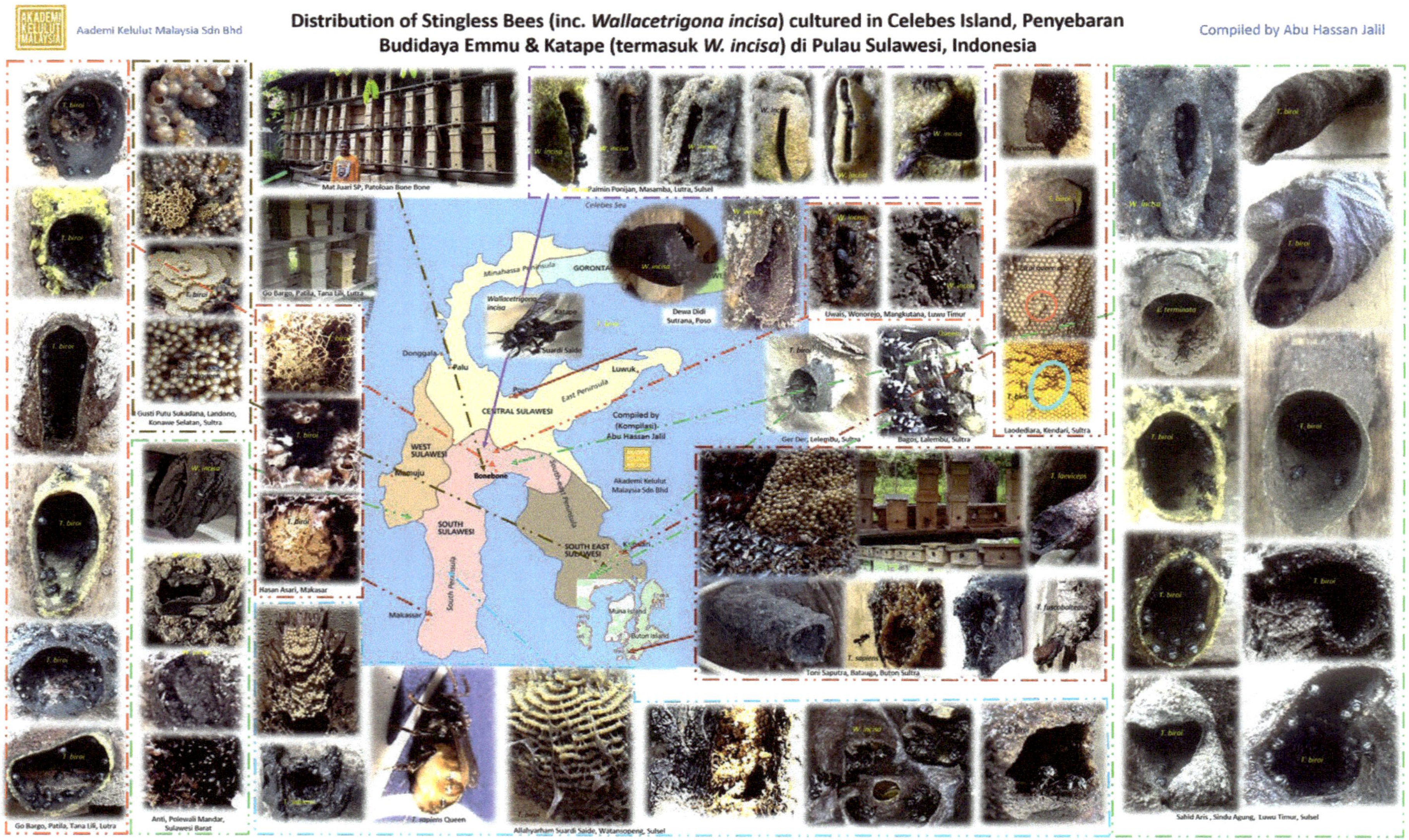

Map 19 Meliponiculture spread throughout Sulawesi

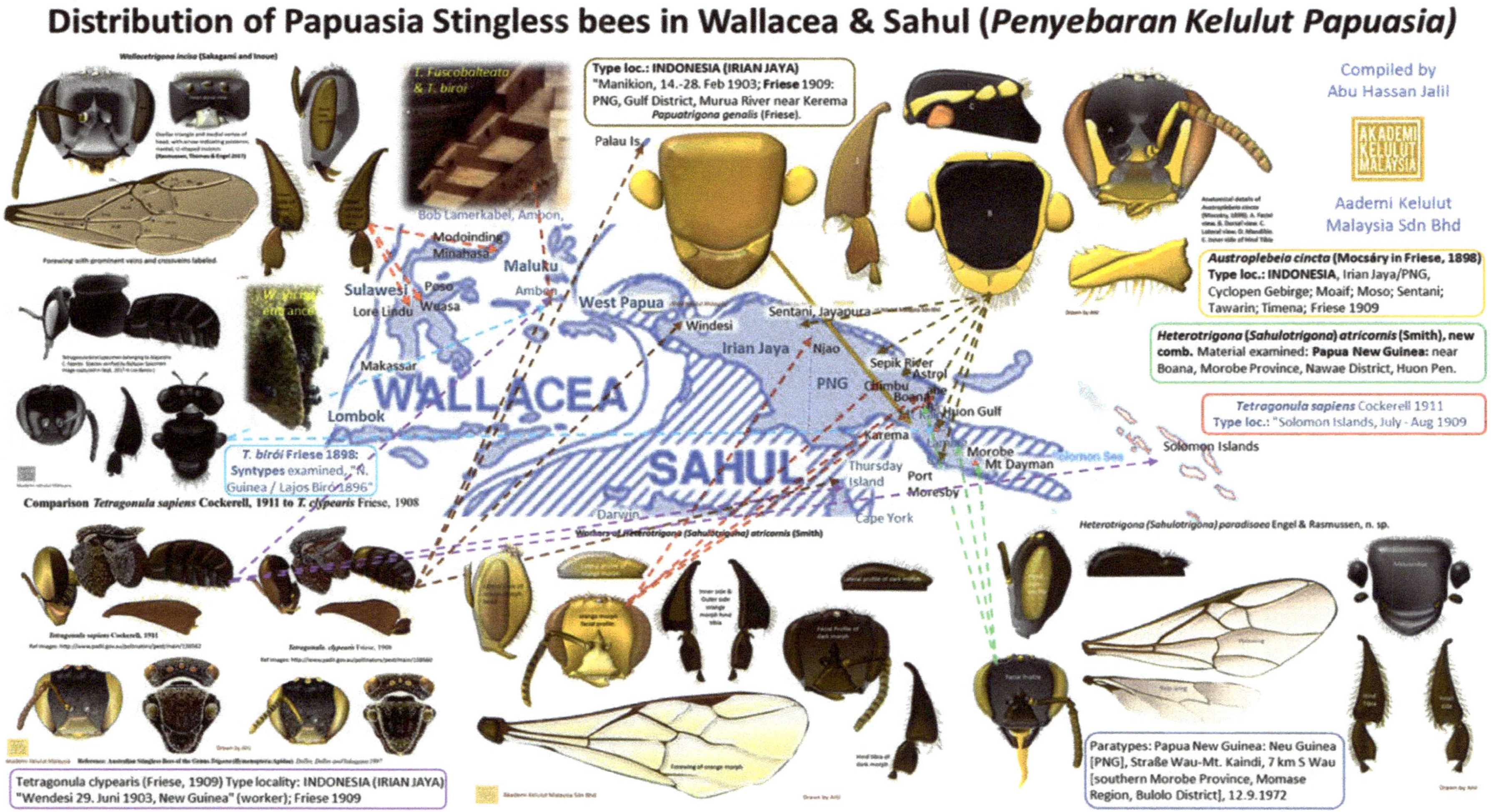

Map 20 Distribution of Papuasia Meliponines in Wallacea & Sahul

Chapter 5-1

5-1.0 - History of Stingless Beekeeping in Wallacea & Sahul.

By Abu Hassan Jalil (abuhsn@gmail.com)

Exploring East of the Wallace Line with the Huxley modification has prompted some investigation in Wallacea, i.e., the gap between the Sahul Shelf and the Sunda Shelf (where Maluku and Sulawesi sit) plus the Philippines. This gap has significance when one observes some Papuan bees' occurrence in the Philippines. *Tetragonula sapiens* of Solomon Island bees are also reported in Lombok, Nusa Tenggara Barat, Indonesia.

Introduction

Upon hearing of the discovery of a new subgenus in Papua and further queries from peers where the name *Sahulotrigona* came about, I set about looking for the derivative taxa. With my keen interest in etymology, I went through my usual avenues and found that the Sahul Shelf proper stretches northwest from Australia through the Timor Sea towards Timor Leste. It also runs from the northern coast of Australia under the Arafura Sea to New Guinea.

"Sahull" or "Sahoel" appeared on 17th-century Dutch maps applied to a submerged sandbank between Australia and Timor. On his 1803 map, Matthew Flinders noted the "Great Sahul Shoal", where Malays came from Makassar to harvest trepang (sea cucumber). The Sahul and Sunda shelves were given their present names by G.A.F. Molengraaff and Max Wilhelm Carl Weber in 1919.
Description Modified map of Maximilian Dörrbecker (Chumwa)
Source=Base map derived from File: Karte von Sunda und Sahul.png. Work of
Map of Sunda and Sahul and the Wallace Line, the Lydekker Line and the Weber Line
Date 1 October 2012, 13:11:25
Author Altaileopard
The much larger Sahul Shelf was suggested in 1845 by G.W. Earl, who called it the "Great Australian Bank".

Sunda & Sahul are underwater continents. However, during the Pleistocene ice age, about 18,000 years ago, the Sahul Shelf was exposed as a whole landmass. The shoreline of this time has been identified in locations that now lie 100 to 140 metres below sea level. Monash University developed a Flash-based interactive timeline of sea-level changes in the 2000s. The Arafura Shelf formed a land bridge between Australia, New Guinea, and the Aru Islands, and these lands share much similar fauna and freshwater fish. Lydekker's Line, a biogeographical line, runs along the edge of Sahul Shelf, where it

drops off into the deep waters of the Wallacea biogeographical area. Wallacea sits in a gap between the Sahul Shelf and the Sunda Shelf, part of the continental shelf of Southeast Asia.
Ref: https://en.wikipedia.org/wiki/Sahul_Shelf accessed 11 February 2018

This segregation later brings a further understanding of the Wallacea and the Sulawesi regions, i.e. east of the Wallace Line. This line bears significance when one considers the new Genus taxa of *Wallacetrigona* for the stingless bee in Sulawesi. Further on, the Wallace Line, with the modification of the Huxley's Line, brings us to the Philippines. The connection comes about when we investigate *Tetragonula biroi* in The Philippines.

Heinrich Friedrich August Karl Ludwig Friese, at the turn of the century, a prolific author for names of bees with more than 2,000 names of species and varieties or subspecies to his credit, authored 16 names of Indo-Australasian stingless bee names. Well diversified in description with patronyms (e.g. *reepeni, biroi, pfeifferi*), names of locality (e.g. *australis, keyensis, borneensis*), anatomical attributes (e.g. *pygmaea, genalis, latipes*), colour attribute (e.g. *flaviventris, luteiventris, versicolor* & etc.). Interestingly, Friese in 1898 named a stingless bee in New Guinea in honour of Lajos Bíró. Type locality in PAPUA NEW GUINEA, Erima. *T. biroi* was named after the outstanding Hungarian collector, Lajos Bíró, who explored the insects of New Guinea from 1896 to 1901 (Pers. Comm. Anne Dollin 2014). [Lajos Bíró (29/8/1856 – 2/9/1931) was a zoologist from Hungary. (Senckenberg Biographies, 2013)]

Dr. Lamerkabel of Universitas Pattimura, Maluku, in his paper on Nesting Characteristics of Stingless Bees in Ambon, Maluku 2017 states that they discovered 282 *T. biroi* colonies in the wilds of seven districts in Ambon. It is no surprise when one considers the occurrence of *T. biroi* also in Davão, Mindanão (Pers. Obs. 2016) right up to Laguna in Luzon, and a specimen verified by Baltazar in Los Baños, Philippines.

This news prompts me to look at the incipience of bees in "the vicinity of Sahul" to include islands East of the Wallace Line with the Huxley modification.

5-1.1 - Stingless beeS of West Halmahera

(Contribution of Hearty Salatnaya)

Beautiful Scenery of Loloda, West Halmahera.

Figure 354 Beautiful Scenery of Loloda, West Halmahera e) f) & g) oval tube entrance hard brittle propolis

I want to share our *Trigona* from West Halmahera. The entrance tube is hard and brittle; the bee has transparent wings and is still unidentified. What is the species?

Abu Hassan Jalil > Hearty Salatnaya, you must be referring to the Maluku Province. From the lateral view (projecting scutellum) of the bee, it is either a *Tetragonula* (small bee uniform wing colour) or *Platytrigona* (medium bee white tip wings). Suppose *Tetragonula* can be found in this area, i.e. *T. clypearis* or *T. sapiens*. However, the nest entrance tube does not look like either of the two. If it is *Platytrigona,* then it could be *P. keyensis* (after key-lander Old Dutch name for Ambon). It could also be a meliponine not identified yet.

Would need clearer close-ups to tell. We do not believe this to be *Tetragonilla collina* because the tube aperture is more oval than round, and the scutellum is not very pronounced.

Chapter 5-2

5-2.0 - Sulawesi

5-2.0.1 - Wallacetrigona incisa behavior and nest condition in Libukang, Sulawesi, Indonesia

(Source: Suardi Saide, English Text by Abu Hassan Jalil)

If this content is a song, this is the part I would sing with much emotion and even maybe teary-eyed. I write this section in memory of a dear old friend, the late Suardi Saide. He spent his last years studying and conserving stingless bees in Sulawesi. I dedicate this section to his family and friends. May God bless his soul, and may his soul rest in peace.

Figure 356 Green Honey from Tetragonula sp. hive

He was the first person in Sulawesi that discovered green honey in his hives.

a) Honey looks green in colour.

b) Looks like a nest entrance covered with green propolis?

c) Honey stocked around the inlet of the nest (appears similar to chunks of an agate-like substance), normally closing the inlet to help protect the colony [these are NOT CHICKENS]

Figure 357 nest entrance G. incīsa, from Day 7 to Day 14 but generally, close the aperture by half later on.

Defective young bee being dumped out. Stingless bees have a very strict protocol for deformed or underdeveloped young.

Figure 359 Defective young bee being dumped out.

Figure 358 Suardi Saide on Wallacetrigona incisa in Libukang, Sulawesi, Indonesia.

Some nests were acquired from the central region, but hundreds are dying not because of the transport process but often caused by entering the wrong nest. It appears that *Tetragonula* sp. can easily beat *Wallacetrigona*.

Figure 360 Wallacetrigona incisa foraging sunflowers

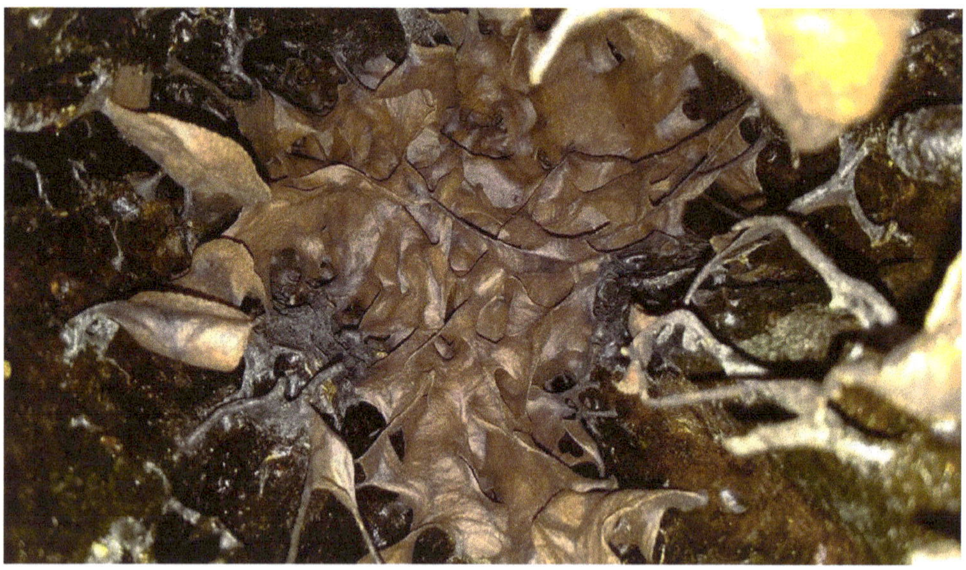

Figure 361 the top of the membrane would be involucrum but the hive seems silent, developments may decline due to high temperatures

Figure 362 G. incisa (Katape), emmu jumbo from Masamba.... G. incisa (katape) and Tetragonula sp. (emmu) Flapping wings to cool the nest

Seems there are some (4) types of pollen brought home

Figure 363 Night mode

5-2.0.2 - Meliponiculture & Bee Plants in Soppeng, South Sulawesi: Photos by Suardi Saide

Mimosa pudica

Cucurbita pepo

Syzygium sp.

Wild flower

Cajanus cajan

Bauhinia blakeana

Ceiba pentandra

Camelia sp.

Celosia spicta

Morus alba

Citrus sp.

Musa paradisiaca

Tithonia diversifolia

White Lotus

Citrus limon

Nelumbo sp.

Nelumbo nucifera

Tradescantia pallida

Lotus

Costus sp.

Syzygium aqueum

Punica granatum

Echinodorus palifolius

Suardi Saide

Figure 364 Bee Plants in Soppeng, South Sulawesi: Photos by Suardi Saide

Figure 366 Meliponiculture & Bee flowers in Soppeng, South Sulawesi: Photos by Suardi Saide

5-2.0.3 - Wallacetrigona incisa queen antenation in Makassar, Sulawesi by Hassan Asri

Figure 365 Wallacetrigona incisa queen antennation in Makassar, Sulawesi by Hassan Asri

Queen antennation is a peculiar behaviour of stingless bee workers surrounding and facing the queen with extended antennae. This phenomenon is not fully understood but occasionally occurs where the antennation lasts for about 10 to 15 minutes, and then everything goes back to normal.

Chapter 5-3

5-3.0 - Papuasian Meliponiculture Potential in Maluku Islands.

By Abu Hassan Jalil

5-3.1 - Ecosystems in Maluku (Province) of Indonesia

Ref: https://en.wikipedia.org/wiki/Maluku_(province) and http://www.terrestrial-biozones.net/Paleotropic%20Ecosystems/Maluku%20Ecosystems.html

Maluku Ecosystems

5-3.1.1 - Maluku Dipterocarp Forest

endemic *Shorea selanica* (Dipterocarpaceae), the only dipterocarp species in the lowland forests of north Seram.

Typical sub-canopy trees include *Calophyllum soulattri*

Map 21 Map of Maluku Islands

5-3.1.2 - Maluku Montane Forests and Summit Formations

On Seram these forests 400 to 1400 m elevation contain *Castanopsis buruana* and *Lithocarpus celebicus*, also *Agathis dammara*, *Coelostegia griffithii* and *Fagraea blumei*. Lower montane zones of Gunung Binaiya have *Syzygium* (possibly *S. aqueum*), conifers such as *Dacrycarpus imbricatus, Phyllocladus hypophyllus, Trimenia papuana* and *Podocarpus characterisa*, along with *Dacrycarpus imbricatus*, up to 35 m high, and shrubs of *Rhamnus*, *Rhododendron* and *Vaccinium* together with the endemic tree ferns *Cyathea binayana* and *C. pukuana* (Cyatheaceae) at 2700-3000 m, endemic herbs — *Euphrasia ceramensis* (Orobanchaceae), *Pterostylis papuanum* var. *ceramensis* (Orchidaceae) and *Viola binayensis* (Violaceae).

5-3.1.3 - Maluku Limestone Forests

Typical tree species include *Gnetum gnemon, Vaticum papuana,* the endemic or near-endemic *Pimeleodendron amboinicum* (Euphorbiaceae), and *Microcos ceramensis* (Tiliaceae) together with species of *Calophyllum, Canarium, Diospyros, Eugenia, Intsia, Litsea, Myristica, Pometia, Pygeum, Saccopetalum* and *Vitex.*

5-3.2.0 - "Maimai Toher" Cultivation Initiative in the Maluku Islands

1. Introduction to the species of "Propolis Bee" Papuasia and Wallacea
2. Needs of food and nest-building materials as well as natural hosts
3. Sources of colony acquisition - Swarm traps and eduction
4. Plant & Flower Cycle - Endemic flowers, local fruits, weather cycle & exotic vegetation
5. Predators, Pests, Pests and Inquilines (Symbionts & Parasites)
6. Box or container housing - Wooden Box, Pottery or Bamboo
7. Maintenance and Revenue
8. Harvest and Post-Harvest
9. Product Packaging and Downstream Product Development
10. Product Processing
11. The Role of Industry Players
12. References and Bibliography

5-3.2.1 - Introduction to the species of Papuasia and Wallacea Meliponines

BEE FAUNA OF NEW GUINEA. Friese 1909

(genera since changed in all species)

A systematic overview of Meliponines

Group *cincta* (thorax shown in yellow).

1. *Tr. cincta* Mocs.

2. *Tr. australis* FRIESE.

Group *carbonaria* (all black).

3. *Tr. carbonaria* SM.

4. *Tr. biròi* Friese

5. *Tr. planifrons* SM.

Group *laeviceps* (abdomen red).

6. *Tr. laeviceps* SM.

7. *Tr. flaviventris* FRIESE.

Groupe *mellipes.*

8. *Tr. mellipes* FRIESE.

9. *Tr. melina* GRIB

10. *Tr. genalis* FRIESE.

11. *Tr. keyensis* FRIESE

12 *Tr. atricornis* SM.

A CLASSIFICATION OF THE BEES OF THE AUSTRALIAN AND SOUTH PACIFIC REGIONS 1965

1. *Trigona (Tetragona) angophorae* Cockerell, 1912b (B.M.)
2. *atrata* (Fabricius, 1798) (Apis) (desc.) Ambon
3. *atricornis* Smith, 1864 (Ox.) New Guinea
4. *biroi* Friese, 1898a New Guinea

5. *carbonaria* Smith, 1854 (B.M.)
6. *clypearis* Friese, 1908 New Guinea
7. *flaviventris* Friese, 1908 (desc.) New Guinea
8. *genalis* Friese, 1908 New Guinea
9. *hockingsi* Cockerell, 1929d (Q.M.)
10. *iridipennis* Smith, 1854 (B.M.) Ceylon to the Solomon Islands
11. *keyensis* Friese, 1901 = *latigena* Friese, 1901 (desc.) Kai
12. *laeviceps* Smith, 1857 (B.M.) Singapore reported from New Guinea and Australia
13. *lamingtonia* Cockerell, 1929c (A.M.) New Guinea
14. *mellipes* Friese, 1898a (desc.)
15. *planifrons* Smith, 1864 (B.M.) New Guinea
16. *sapiens* Cockerell, 191 lb Solomon Islands
17. *wybenica* Cockerell, 1929a, 1929d (U.S.N.M.)

UPDATED CHECKLIST OF STINGLESS BEES (MELIPONINI) FROM PAPUASIA 2019

Genus *Austroplebeia* Moure, 1961

Austroplebeia cincta (Mocary in Friese, 1898)

Genus *Heterotrigona* Schwarz, 1939

Subgenus *Platytrigona* Moure, 1961

Heterotrigona (Platytrigona) flaviventris (Friese, 1908 [1909])

Heterotrigona (Platytrigona) keyensis (Friese, 1901)

Heterotrigona (Platytrigona) lamingtonia (Cockerell, 1929)

Heterotrigona (Platytrigona) planifrons (Smith, 1865)

Subgenus *Sahulotrigona* Engel & Rasmussen, 2017

Heterotrigona (Sahulotrigona) paradisaea Engel & Rasmussen, 2017

Heterotrigona (Sahulotrigona) taraxis, n. sp.

Heterotrigona (Sahulotrigona) tricholoma, n. sp.

Genus *Papuatrigona* Michener & Sakagami, 1990

Papuatrigona atricornis (Smith, 1865), n. comb.

Genus *Tetragonula* Moure, 1961

Subgenus *Tetragonula* Moure, 1961*

Tetragonula (Tetragonula) biroi (Friese, 1898)

Tetragonula (Tetragonula) clypearis (Friese, 1909)

Tetragonula (Tetragonula) sapiens (Cockerell, 1911)

5-3.3.0 - Needs of food and nest-building materials as well as natural hosts

- **Nektar & Pollen**
- **Floral Records**: Angophora; *Antigonon leptopus; Asystasia gangetica; Callistemon citrinus; Coccinia; Cocos nucifera; Dillenia suffruticosa; Euphorbia pulcherrima; Galphimia glauca; Hamelia patens; Impatiens balsamina; Ixora; Jatropha hastata; Jatropha podagrica; Leucaena leucocephala; Melastoma; Mussaenda; Nerine; Nicotiana tabacum; Plumbago auriculata; Canna flavescens*
- **Crops**: Kisa *(Citrus sp.),* Koli palm *(Arecaceae)* orange, Corn *(Zea mays L.) (Cucurbitaceae)* and cassava *(Manihot esculenta L. Crantz),* Groundnuts *(Arachis spp.)* and beans *(Phaseolus spp.)* lesser yams *(Dioscorea esculenta Lour. Burkill); Cucumis sativus; Lycopersicon esculentum*
- **Swamp**: *Rhizophora apiculata*; *Cupaniopsis anacardioides*

- **Gum, Sap & Resinous Trees**
- **Tree visitations records:** *Abrophyllum ornans; Acradenia euodiiformis; Alphitonia excelsa; Euroschinus falcata; Rhodomyrtus psidioides; Tristaniopsis laurina; Pollia Crispata; Eucalyptus staunus; Alphitonia; Eucalyptus; Melaleuca; Rapistrum; Swainsona; Wahlenbergia*; *Santiria laevigata; Banksia; Callistemon; Claoxylon; Cupaniopsis; Daviesia; Dendrobium; Eucalyptus; Hibbertia; Jacksonia; Leptospermum; Dendrobium teretifolium; Cycas; Cymbidium madidum;*
- **seed dispersal** *Dendrobium toressae; Eucalyptus* torelliana; *(***mellitochory**);
- **Crops;** Breadfruit *(Artocarpus sp.);* Cloves *Syzygium aromaticum;* Nutmeg *Myristica fragrans; Macadamia integrifolia*

5-3.3.1 - Ensuring food security in the small islands of Maluku: A community genebank approach (Semuel Leunufna & Mike Evans).

Some genetic resources are found in the Kisar, Leti and Moa islands South West South Maluku Regency of the Maluku Province. Kisa (*Citrus* sp.), Koli palm (Arecaceae) and Orange are considered germplasm specific to the region. Corn (*Zea mays* L.) is the main food crop, grown in multiple cropping systems with a legume, cucurbits (Cucurbitaceae) and cassava (*Manihot esculenta* L. Crantz) are mostly attacked by grasshoppers. Groundnuts (*Arachis* spp.) and beans (*Phaseolus* spp.) are also cultivated, while lesser yams (*Dioscorea esculenta* Lour. Burkill) are mainly grown wild, uncultivated.

Two different perennial food crops, such as breadfruit (*Artocarpus*), are also found. Maluku Native Forest Clove (*Syzygium aromaticum* L. Merr & Perry), and Moa buffalo (*Bubalus* sp.) native to Moa Island.

5-3.4.1 - Sources of colony acquisition - Swarm traps

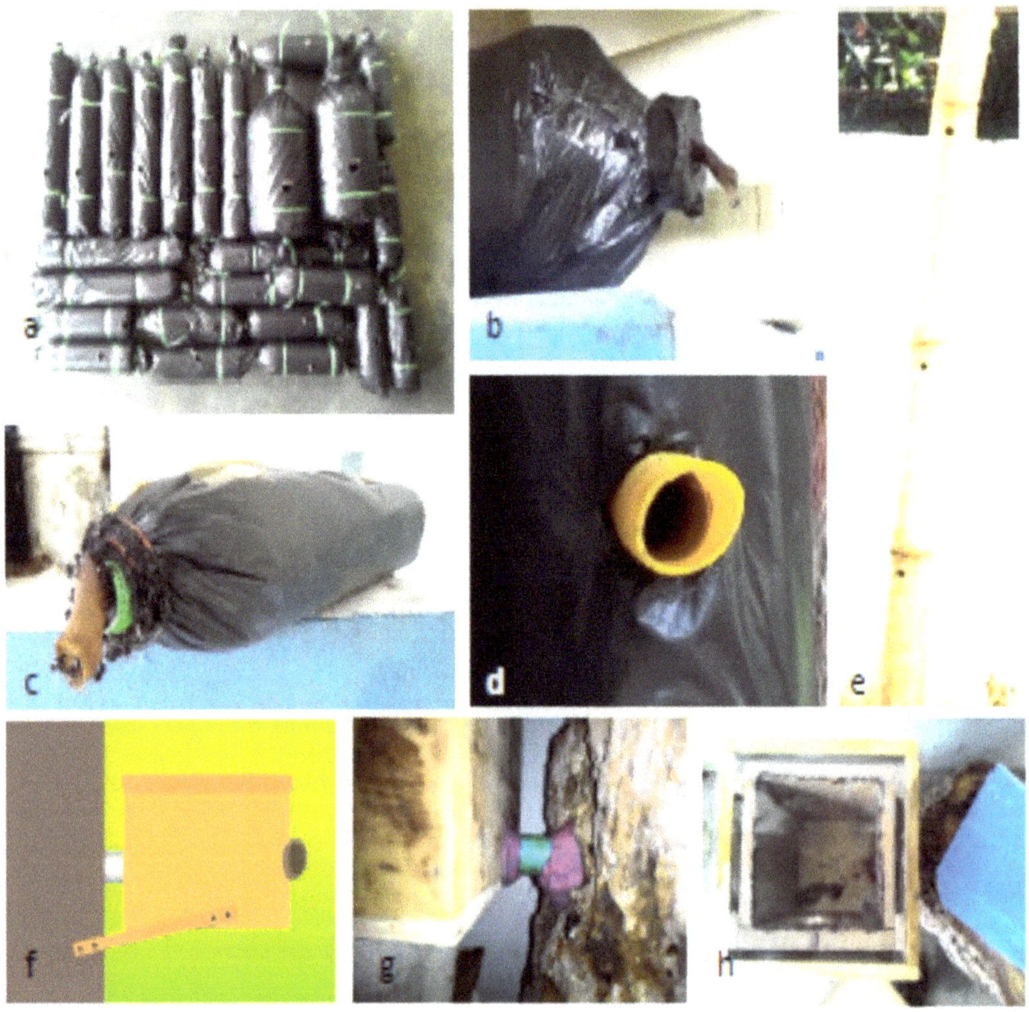

Figure 367 Colony acquisition by swarm traps and eduction

5-3.4.2 - Sources of colony acquisition – Eduction

Figure 368 Eduction Methods

5-3.5 - Predators, Enemies, Pests and Inquilines (Symbionts & Parasites)

5-3.5.1 Hive & colony pest

(*Achroea grisella:* Tineidae: Lepidoptera)*;* (*Protaetia mandarinea:* Scarabaeidae: Coleoptera), (*Brachypeplus auritus*: Nitidulidae: Coleoptera)

Figure 369 Hive & colony pest

5-3.5.2 - Predator & Enemies (Insects)

(Bembix musca: Crabronidae: Hymenoptera)); (Dorniphora trigonae: Phoridae: Diptera) (of brood?)); Small Hive Beetles *Aethina tumidus*

Figure 370 Predator & Enemies (Insects)

5-3.5.3 - Inquilines

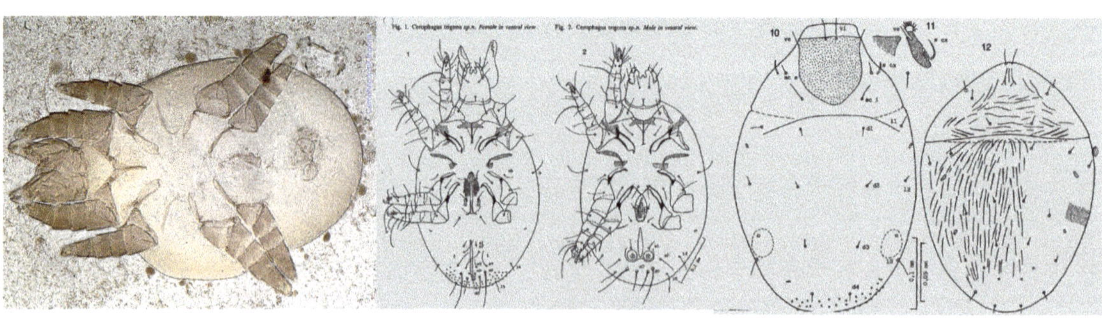

Figure 371 (Inquiline mite (Cerophagus trigona: Acaridae: Acari)

(*Brachypeplus basali, B. meyricki* (or *Epuraea), B. planus;* Nitidulidae: Coleoptera)*;* (*Tribolium myrmecophilum:* Tenebrionidae: Coleoptera*)*; (*Homalota trigonae:* Staphylinidae*:* Coleoptera*)*; Mites (*Cerophagus trigona*: Acaridae: Acari);

5-3.5.4 - Predator Mammals

Anteaters, Pangolins, Primates, Cuscus

Ornate Moluccan **cuscus** *Phalanger ornatus* (Gray, 1860)

Ornate cuscus range

The ornate cuscus or Moluccan cuscus (*Phalanger ornatus*) is a species of marsupial in the family Phalangeridae. It is endemic to Indonesia, where it is found on the North Maluku islands of Halmahera, Bacan and Morotai, at elevations from sea level to 1000 m.

Figure 372 Ornate Molluccan cuscus

https://en.wikipedia.org/wiki/Ornate_cuscus

Groves, C. P. (2005). "Order Diprotodontia". In Wilson, D. E.; Reeder, D. M (eds.). Mammal Species of the World: A Taxonomic and Geographic Reference (3rd ed.). Johns Hopkins University Press. p. 47. ISBN 978-0-8018-8221-0. OCLC 62265494.

^ Jump up to [a] [b] *Leary, T.; Singadan, R.; Menzies, J.; Helgen, K.; Wright, D.; Allison, A. & Flannery, T. (2008). "Phalanger ornatus". IUCN Red List of Threatened Species. **2008**. Retrieved 28 December 2008.*

5-3.5.5 - Predator (small birds)

Beefeaters, Honeyeaters, Kingfishers, Flycatchers, Woodpeckers. Insectivorous birds

Figure 374 Small Birds – Bee-eaters (Meropidae)

Figure 373 Sulawesi and Halmahera: Birding the Edge of Wallace's Line

5-3.5.6 - Predator Reptiles, Amphibians, Arachnids

https://www.inaturalist.org/projects/insects-and-spiders-of-north-maluku

Figure 375 Lizards, Frogs, Spiders of N. Maluku

5-3.6 - In Summary

The practice of Meliponiculture with a conservation objective can be beneficial to the socio-economic status of farmers and communities at the forest edges.

Maintenance & Income

A regular maintenance regime in meliponary management is of utmost importance. The regime comes in daily, weekly and monthly activities concerning the flowering, fruiting and climatic season cycles. Provision and upkeep are recommended for foraging resources of vegetation, nest-building materials and other water, salts and essential minerals.

Colony placement and density in the bee farm should be well planned, with a transit corner provided that can serve as an incoming section for newly acquired colonies that may also serve as a sickbay [but the diseases are transmitted at flowers or other resources, so what is the purpose- to prevent parasite spillover?] when the need arises.

Livestock farm well segregated from the bee farm for Permaculture practices. Ambient parameters control temperature, humidity and strong prevailing winds.

Pest control of insect predators & other animal predators is scheduled routinely.

Income is very much dependent on the following issues.

1. Honey Harvest Routine
2. According to the season
3. Scheduled
4. Weather suitability
5. Propolis Harvesting Routine and Bee Bread
6. Depending on demand
7. Colony Maturity
8. Quantity or Most resources instead

5-3.7 - References for Papuasian Meliponiculture

1. Die Bienenfauna Von Neu-guinea Von Dr. H. Friese. Annales Musei Nationalis Hüngarici. 1909
2. Catalogue of the Indo-Malayan/Australasian stingless bees (Hymenoptera: Apidae: Meliponini) Claus Rasmussen 2008 Zootaxa
3. Morphological Traits of Maluku Native Forest Clove (*Syzygium aromaticum* L. Merr & Perry) Article in Journal of Tropical Crop Science · July 2019
4. A Key to The Genera And Subgenera Of Stingless Bees In Indonesia (Hymenoptera: Apidae) Michael S. Engel, Sih Kahono and Djunijanti Peggie. Treubia 45: 65–84, December 2018
5. A new subgenus of *Heterotrigona* from New Guinea (Hymenoptera: Apidae) Journal of Melittology November 2017. Claus Rasmussen
6. Ensuring food security in the small islands of Maluku: A community genebank approach Semuel Leunufna a, Mike Evans b,c, Journal of Marine and Island Cultures 2014
7. *Aleurites moluccana* (L.) Willd. Ecology, silviculture and productivity Haruni Krisnawati, Maarit Kallio & Markku Kanninen. CIFOR 2011 Center for International Forestry Research
8. Description and life cycle of *Cerophagus trigona* spec. nov. (Acari, Acaridae), associated with the stingless bee *Trigona carbonaria* SMITH In Australia By A. FAIN And T.A. Heard ENTOMOLOGIE, 57: 197-202, 1987
9. Notes on Papuasian and Malesian stingless bees, with the descriptions of new taxa (Hymenoptera: Apidae) Michael S. Engel Journal of Melittology 2019
10. Current generic classification of sap beetles (Coleoptera, Nitidulidae) Alexander Kirejtshuk Zoological Institute, St. Petersburg, 2008
11. A Classification of The Bees Of The Australian And South Pacific Regions Charles D. Michener Bulletin Of The AMNH VOL 130 NY 1965

Chapter 5-4

5-4.0 - Meliponiculture Activities in West Papua.

By Edhi Yansyah 13 June 2021

Figure 376 Meliponiculture efforts by Edhi Yanshah in Misool

Raja Ampat also has stingless bees. So, the story is that last week, we wanted to cultivate trigona honey bees (stingless bees) in Lalelkai Misool. When we put the bee setup box, we just brought it from Sorong. It turned out that a local species was nesting in the foundation of the land house. They attacked the bees we brought. We had to move the bees from outside to a safe place not to be attacked by the local bees in Lalelkai.

Figure 377 Lokasi budidaya lebah trigona (lebah tdk menyengat/klanceng) di Lalelkai Misool.

Figure 378 Tetragonula cf laeviceps found in rock fissures.

Figure 379 The colony we dismantled from the last rock without a queen already has a queen ... Development in the hive setup.

5.4.1 Online chat with Edhi Yanshah, LaHisa & AHJ via WhatsApp

[12:06, 12/11/2021] Edhi Yansyah: Ready ... I am very grateful to Pak Abu; I almost bought the itama colony ... Even though there is also a large Kelulut in Papua...

Earlier I saw that some were collecting sap. I already told the children who guard the island to find their nest.

Figure 380 This is our island, separate from the other islands around ... Maybe if I could have a collection of Sahul types on the island, it could be Sahul Meliponini Park

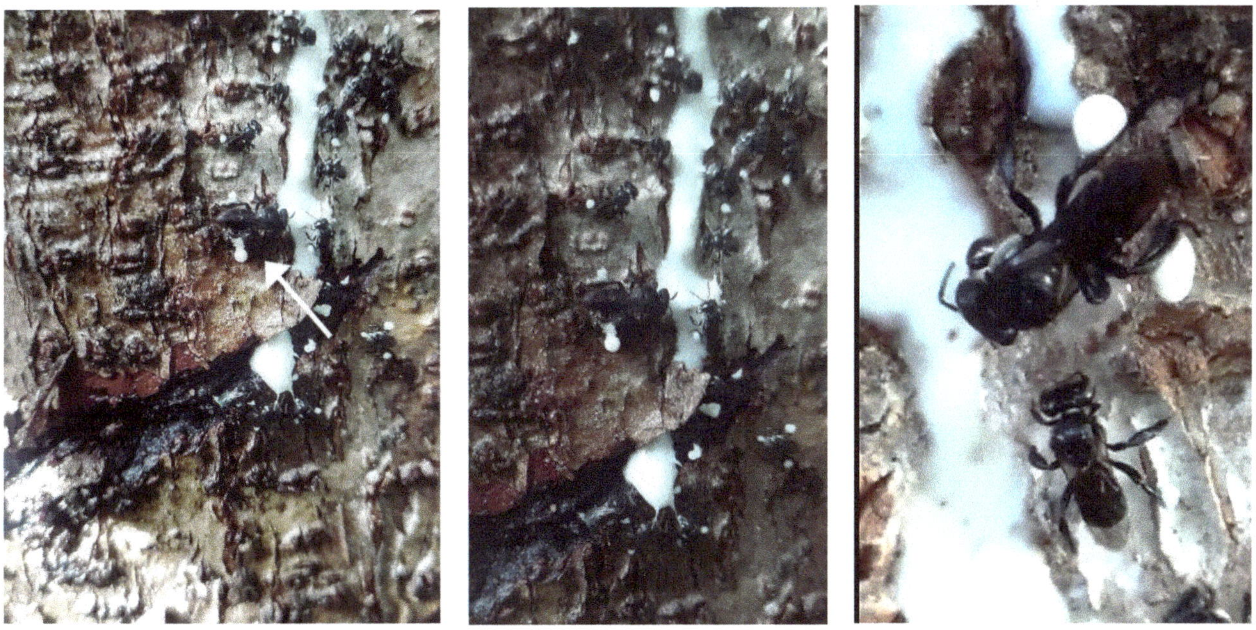

Figure 381 A large size stingless bee relative to the common smaller ones. Suspected to be Sahulotrigona paradisaea.

Figure 382 Agnez & Fiska are school teachers and just 6 months started Meliponiculture in Yanggandur, Sota, Merauke, West Papua.

[09:40, 10/12/2021] Edhi Yansyah: Wow, it's easier to be invited to cultivate kelulut ... I in Raja Ampat still don't have many who want it. They just heard it before moving.

[08:43, 10/12/2021] LaHisa Merauke: If in the next village (Marori tribe), the area's name is Ndokof, this is for the mention of the two species of bees. If Agnes's mother is in the Kanum tribe. Even though the two tribes are only +/- 30 km apart, the languages are different

[08:36, 10/12/2021] +62 823-9880-2019: *Mentya* (*Tetragonula mellipes*) and *Bress* (*Austroplebeia cincta*)

Yes, the language of the Kanume tribe is one of the tribes in Merauke Regency (South Papua)

[09:25, 10/12/2021] Edhi Yansyah: *Bebiyem* if the language is Biak ... It will be very difficult to use regional languages in Papua because Papua has a variety of languages that are almost the same as the number of regional languages in Indonesia

Map 22 Yanggandur, Sota, Merauke, West Papua.

[09:28, 10/12/2021] AHJ: Thank you, Ibu Agnes, Pak LaHisa and Mr. Edhi. This list creates a collection of vernacular names for researchers on an expedition to the interior.

[09:35, 10/12/2021] LaHisa Merauke: Ibu Fiska entered this activity and started kelulut bees, one village with Ibu Agnes.

Ibu Agnes & Ibu Fiska are school teachers and started Meliponiculture in Yanggandur, Sota, Merauke, West Papua for just six months.

5.4.2 Colonies were found in termite nests

*Figure 383 Colonies were found in termite nests (*Macrotermes sp*?) up to +/- 3 m in savannah and open woodland forests in southern Papua (Wasur National Park); the colony occupies the cavities in the termite nest;*

Colonies were found in termite nests (*Macrotermes* sp?) up to +/- 3 m in savannah and open woodland forests in southern Papua (Wasur National Park); the colony occupies the cavities in the termite nest; to get the colony, we carve the termite nest using a machete, axe or pointy wood; eggs and colonies are transferred to boxes, the boxes are placed in the former colony cavity for 1 or 2 days so that all colonies are included in the box; if all the colonies have entered, can be brought home at night.

I got the colony on 6 November 2021. After the colony was put into the box, I moved it to a cultivation area in the forest, too, but it's easy to monitor, so they don't feel much environmental change. Even though the colony is in an artificial box, they still feel the forest environment. Two weeks ago, I went to have a look, the hive structure has recovered, and I started collecting some honey and pollen bags

Figure 384 Images of Heterotrigona (Platytrigone) *cf.* flaviventris *(Friese, 1908[1909]) nest from Bpk La Hisa, Wasur, Merauke*

Figure 385 Austroplebeia cincta of Merauke, West Papua

5.4.3 Online chat between AHJ & Ms Fiska via FB Messenger

AHJ: How long have you been involved in Meliponiculture?

Fiska: Still new, sir. Around the 10th month of 2021. My model is from YouTube, and I'm desperate to try 16 colonies. It turns out that many have run away, sir. It turns out that my setup is too big.

AHJ: May I know a little bit about yourself?

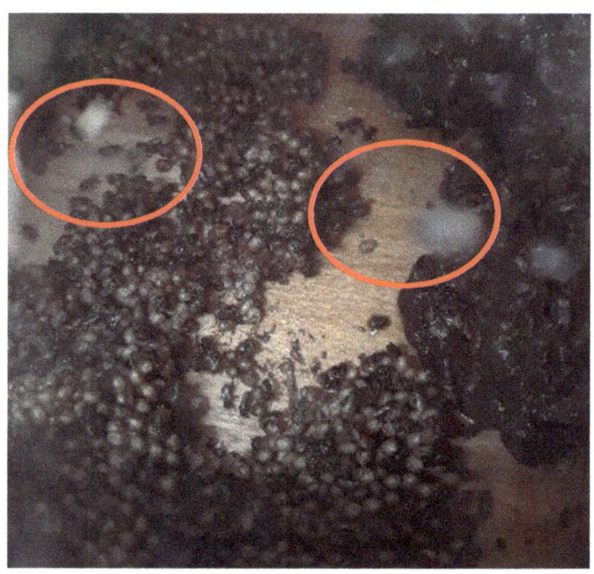

Fiska: I am 31 years old, married with teo children. My name is Fiska Desmawati, from Yanggandur, Merauke.

AHJ: Are familiar with Pak LaHisa?

Fiska: That's right, I called him uncle LaHisa. At the same time, we are trying a new setup that uncle LaHisa brought. Not long after being transferred to the setup, the spilt honey turns into fungus, sir, is there no need for the honey to be added to the setup when transferring the colony?

Figure 386 Tetragonula cf mellipes salvaged from the Termite mound.

That is called *Mentya* by the locals here. Not long after transferring, the remnant spilt honey aggravated fungus growth.

AHJ: Yes, there is no need to add in the honey when transferring

Fiska: Thanks; I want to refill the setup whose colonies have flown away[2]. Permission to ask again, sir. For lizard pests, is there a solution, sir? I put the setup on the pillar, but the lizard always climbs up, and there are two setups that the lizard enters until the colony runs out.

AHJ: For boxes with blurry spots (mycelium), you need to 'roast' over flame or smoke first so that it can remove fungus spores if you want to reuse them. As for lizard traps, many online stores sell commercial lizard traps. You can try an assortment of them e.g., https://shopee.co.id/Perangkap-Jebakan-Cicak-dan-Kecoa-Efektif-untuk-semua-Area-Lingkungan-i.92821435.3637350393?sp_atk=b10b59bd-27d9-4264-ab70-80ec3e8bde6c

Fiska: Yes, sir, I will try that. Let me ask again, sir. The plastic placed under the box cover for the observation should be glued with what, sir?

AHJ: double-sided tape or simply staple.

[2] Editor's note: Colonies do not 'disappear or abscond', generally, they slowly die, especially in a meliponary where there is no mother colony nearby (in nature, there is ALWAYS a good chance of re-connecting to the mother colony, or even a sister colOny). As to meliponary bubbles, I fully agree, and it is an entirely artificial habitat, with some good things for some species, but mostly not much, in comparison to real forest, where there are usually heightened dangers (and rewards).

Fiska: This is my Bee House beside my house.

Fiska: On the left of the yellow flower is *Mentya*. If the one on the right on a red flower has a little yellow pattern on its head, it's called *Bress* by the people here. The cell phone camera is so low-focused.

AHJ: It’s alright, thanks for sharing; maybe next time you get a better opportunity. I also want to see the shape of the roof of this bee house structure from afar.

Fiska: This is the box I ordered at the furniture factory; based on the picture I made, I thought the bigger the place, the more freely the bees would have. It turns out that I was wrong, and 16 setup models like that have already been made. But some are still holding on, sir. Only in the front compartment, and no bee has penetrated the back compartment.

AHJ: Boxes that are too large create problems for bees to regulate their ideal temperature.Your label - Madu Trigona SENT… what is meant by sent?

Fiska: SENT uses the local language here, which means honey. The local language is the Kanume language. I entered the photo that I asked for from Uncle Laisa to clarify the type. Please correct me if something is wrong or not right, sir.

AHJ: Yes, it's good to use the regional language. Only Trigona is not a regional language. Even in Australia, the beekeepers are already using *Tetragonula* & *Austroplebeia* or just a sugar bag bee. Don't use Trigona anymore because it has been found that it is a genus only found in the Americas, not in our region here. You probably printed many labels, and it will be costly to change.

Fiska: I have changed it, sir. Fortunately, the previous one was only temporary. Finally, I updated the writing on the label. And after I entered the chat group, thank God, someone responded, answered and gave a lot of input.

Bit by bit, I will make corrections, thank

Chapter 5-5

5-5.0 - Meliponiculture in Yanggandur Village, Merauke Papua

By Theresia Agnesia Maturbongs

5-5.1 Introduction

I am Theresia Agnesia Maturbongs, and people call me Agnes or Hesty. I am 37 years old, and I work as a teacher in a small junior High School in Yanggandur Village, Merauke Papua. Yanggandur village is about 68 km from Merauke City. The citizens here are the original Kanume tribe. Kanume is one of the other tribes in Southern Papua. Yanggandur village lies in the middle of the forest, and it is a part of Wasur National Park. Every day I teach my students at school. After school, I take care of my stingless bee in the hive. I love and am interested in my new world. My new passion is the stingless bee. I said new world because I am a beginner, and this is something new that I should try to learn much about them. I am a beginner, so I need support and suggestion to increase the production of stingless honey. So, I try to take care of them in the hives.

5-5.2 Background

Since early 2019, I started to product Pure Honey of Stingless Beez. Firstly, I worked alone, then two mama/women Papuan also helped me. The Stingless Bee in this Village has a harvest time. The harvest time of stingless bees is started at the end of February until July. Maybe it is affected by the population of wildflowers and wild plants in the jungle. At this time, there are so many mayflowers in the jungle. They can support resin, nectar, and pollen needed by the stingless bee. Stingless bees here live in an old tree, in ***Musamus*** or people said the ant's nest. No one person is trying to keep them in the hives. If they want honey, they will go to the jungle, cut the tree using an axe and take the honey. Then, in my perception, if we take them by using the way above, it is not effective, and the stingless bee will go extinct, and in the future, our grandchild does not know what stingless bee is or how it looks like and so on. So, I decided to keep them in the hives. I am not an expert in the insect sector, but I want to learn and try it. If there is an effort, I can find the way.

Figure 388 The mechanism of packaging madu pokos yanggandur

Figure 387 Left: Bress Honey with its propolis; Middle: Mentya Honey with its propolis; Right Branding Product & Packaging assortment

5-5.3 The Mechanism of Packaging Madu Pokos Yanggandur

For the first time, we received honey from the jungle with propolis. Then we strained twice by strainer manually. Then we put it into the plastic box dispenser. Then we measured it by using a scale. After that, we put it in the bottles and did not brand our product.

Figure 389 Left: Mentya Madu Pokos 300gr; Middle: Bress Madu Pokos 200gr; Right: Madu Pokos Advert

At the end of 2020, we tried to brand our product and use the new packaging. Maybe people will ask us, where do we get the honey? Is it from the forest or the imitation hives? For the first time, pure honey came from people in this village. I chose this stingless bee because it is very kind to the environment and many populations. There are so many old trees that become stingless bee nests. Sometimes, the trees fell, and the honey spread out everywhere. There is also *cerana indica* bee and *Apis dorsata*, but I prefer to pet Stingless Bee.

Figure 390 Newly harvested honey Initial setup beside house.

Many Stingless Bee colonies in this village live wild in the forest. There are two kinds of stingless bees, i.e., Bress (honey from *Austroplebeia cincta*) and Mentya (honey from *Tetragonula* cf. *mellipes*. Bress and Mentya, I adapted from Kanume Language (Kanume: a local origin tribe in Southern Papua). Bress Honey has a sweet taste; on the contrary, Mentya Honey has a little acid and sweet taste.

Figure 392 Left: Mentya Hive (Tetragonula cf. mellipes); Middle: Bress (Austroplebeia cincta); Right: The entrance of Austroplebeia Cincta

Figure 391 Left: Nest structure of Austroplebeia; Middle: The entrance of T. *cf.* mellipes; *Right: The" Bevak "is a shack for stingless bee*

Austroplebeia cincta and *Tetragonula* cf. *mellipes* are different. *Austroplebeia* has an entrance, but *Tetragonula* cf. m*ellipes* does not have a long entrance outside. It is just a short entry. The colour is dark brown, and they made it from inside.

They always look for the flowers of Bus Tree, Acacia, Eucalyptus Tree and so on as their food. Also, many kinds of Wild Flowers in the jungle. I tried to take care of them in the hive to make plants vegetation, and I planted some kinds of flowers to improve the sources of their resin etc. I put them in a small shack called "Bevak" to keep them from extreme weather and ultraviolet. Bevak is made of wood, and the roof is made of tree bark to make them comfortable. Besides Bevak, I put them under the roof and in a paranet to avoid harmful ultraviolet radiation.

5-5.4 The Obstacles

The obstacles that I found are the limited knowledge about these bees and taking care efficiently of the bees. Some colonies are lost and leave their hives. The other obstacle is we have limited tools for increasing the production of the honey bee. So, I should learn more about stingless bees. I just watched YouTube Channel and so on to get more information about them. In December 2021, I thanked other experts in this case and joined their group. I am really happy to get a new friend and new information. One day I hope if there is training about how to take care of the stingless bee, I want to join it to enrich my knowledge about them.

5-5.5 Marketing of Madu Pokos Yanggandur

The Animo of Merauke Citizens to Madu Pokos is good because they prefer raw honey without pasteurization. Mostly our marketing is by online social media such as Facebook, WhatsApp, Instagram, Tiktok etc. The enthusiasm of the Merauke people is really good, but we always have limited or empty stock. To solve this problem, I tried to keep them in the hive, and we have transferred colonies from a log into new hives. I hope Madu Pokos Yanggandur is well known in Indonesia and other people abroad one day.

Figure 393 Where there are bees, there are flowers, and where there are flowers, there is new life and hope

5-6 References for Section 5

Ascher, J., & Pickering. (28 Feb 2018). *Discover Life bee species guide and world checklist.* Retrieved from Sahulotrigona: http://www.discoverlife.org/mp/20p?see=Sahulotrigona&guide=1&flags=HAS:subgenus:

Ascher, J., & Pickering. (28 Feb 2018). *Discover Life bee species guide and world checklist.* Retrieved from Austroplebeia: http://www.discoverlife.org/mp/20q?search=Austroplebeia&guide=1&flags=HAS:subgenus:

Ascher, J., & Pickering. (28 Feb 2018). *Discover Life bee species guide and world checklist.* Retrieved from Wallacetrigona: http://www.discoverlife.org/mp/20q?search=Wallacetrigona&guide=1&flags=HAS:subgenus

Ascher, J., & Pickering. (29 Feb 2018). *Discover Life bee species guide and world checklist.* Retrieved from Tetragonula: http://www.discoverlife.org/mp/20q?search=Tetragonula&guide=1&flags=HAS:subgenus:

Engel, M. S., & Rasmussen, C. (2017). A new subgenus of Heterotrigona from New Guinea (Hymenoptera: Apidae). *Journal of Melittology.*

FREISE, H. (1898). Die Trigona-arten Australiens. *Termeszetrajzi Fuzetek 21:*, 427-431.

Goyret, J., & Farina, W. M. (2003). A descriptive study of antennation during trophallactic unloading contacts in honeybees Apis mellifera carnica. *Insect. Soc. 50 (2003) 274–276.*

Jalil, A. (2014). *Beescape for Meliponines.* Singapore: Partridge Publishing.

Jalil, A. (2016). *Meliponine Identifier Pictoruial Guide.* Akademi Kelulut Malaysia.

Lamerkabel, J. S., & et al. (2017). Tabiat Bersarang Dua Spesis Stingless Bee Koloni Di Pulau Ambon Muluku. *Seminar Nasional Perlebahan.* Bogor: IPB.

Moure, I. (1961). Preliminary Supra-specific classification of the old world Melliponinae Bees (Hymenoptera, Apoidea). . *Studia Entomology. 4:* 181-242.

Rasmussen, C. (2008). Catalogue of the Indo-Malayan/Australasian stingless bees (Hymenoptera: Apidae: Meliponini). *Zootaxa 1935.*

Rasmussen, C., & Camargo, J. M. (2008). A molecular phylogeny and the evolution of nest architecture and behavior in Trigona s.s. (Hymenoptera: Apidae: Meliponini). *Apidologie 39.*

Rasmussen, C., Thomas, J. C., & Engel3, M. S. (2017). A New Genus of Eastern Hemisphere Stingless Bees (Hymenoptera: Apidae), with a Key to the Supraspecific Groups of Indomalayan and Australasian Meliponini. *AMERICAN MUSEUM NOVITATES.*

S.D., L., Zeilhofer, S., Blu¨ thgen, N., & Schmitt, T. (2010). Stingless Bees Use Terpenes as Olfactory Cues to Find Resin Sources. . *Chem. Senses 35:* , 603-611.

Sakagami, S. F., Inoue, T., Yamane, S., & Salmah, S. (1983). Nest architecture and colony composition of the Sumatran stingless bee Trigona (Tetragona) laeviceps. . *Kontyu, 51(1):*, 100-111.

Sayustia, T., Raffiuddin, R., Kahono, S., & Nagir, T. (2021). Stingless bees (Hymenoptera: Apidae) in South and West Sulawesi, Indonesia: morphology, nest structure, and molecular characteristics. *Journal of Apicultural Research Volume 60, 2021 - Issue 1.*

Schwarz, H. F. (1937). Bornean Stingless Bees Of The Genus Trigona. *Bulletin American Museum of Natural History.*

Schwarz, H. F. (1939). The Indo Malayan Species of Trigona. *Bulletin of AMNH.*

Syafrizal, F., Tarigan, D., & Yusuf, R. (2014). Biodiversity and Habitat of Trigona at Secondary Tropical Rain Forest of Lempake Education Forest, Samarinda, Kalimantan Timur. *Jurnal Teknologi Pertanian 9(1):34-38,.*

Trevero, J. T., Nuez, I. L., & Sagliba, W. S. (2012). Pollination of Apis mellifera and Trigona biroi. *IJERD 3-2.*

Section 6 – Beyond Meliponiculture

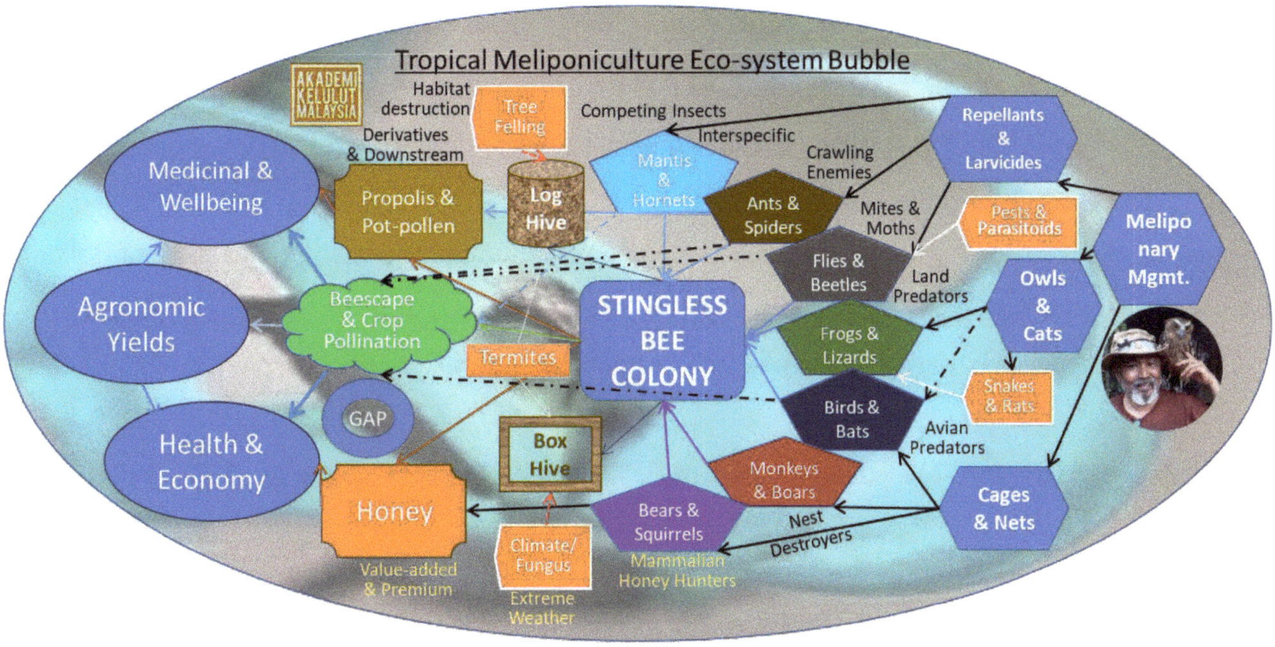

Contents in Section 6

Chapter 6 -1

6-1.0 - Additional Nest Structures and Natural Enemies of Stingless Bees

(Hymenoptera; Apidae; Meliponinae).

Norita Widya Pangestika[1] (noritawidyapangestika@gmail.com), Tri Atmowidi[2] (atmowidi@gmail.com) & Sih Kahono[3] (sihkahono@gmail.com)

[1]Animal Bioscience, Post Graduate School, Bogor Agricultural, University, IPB Dramaga Bogor 16680.

[2]Department of Biology, Faculty of Mathematics and Natural Sciences, IPB Dramaga Bogor 16680.

[3]Zoology Division (Museum Zoologicum Bogoriense), Research Centre for Biology- LIPI, Cibinong 16912.

ABSTRACT

Stingless bees (Hymenoptera: Apidae) are widely distributed in the tropics and certain subtropical areas. Humans farm these bees because they produce honey and other products. This study aimed to study the common nest structure and additional nest structures of stingless bees that may be caused by natural enemies that frequently attack them. This research was conducted from August to November 2015. Samples were taken from three locations: Pasuruan (East Java), Tasikmalaya (West Java), and Rangkasbitung (Banten). Descriptive analysis was used to compare the structure and composition of the nests in bamboo, concerning natural enemies.

Generally, the structure and composition of the nest of stingless bees in bamboo consist of the nest entrance, storage pots (honey and pollen), and brood cells. We didn't find natural enemies in Pasuruan and Tasikmalaya. These colonies did not create additional nest structures. In contrast, the colonies in Rangkasbitung created additional nest structures, i.e., the inner entrance covered by batumen and cerumen and built a root-like structure. Natural enemies found in the colony of Rangkasbitung were cockroaches (Blattodea: Blattidae), histerid beetles, *Platysoma leconti* (Coleoptera: Histeridae), nitidulid beetles, *Carphophilus* sp. (Coleoptera: Nitidulidae), and ants, *Pheidole* sp. (Hymenoptera: Formicidae).

Keywords: Stingless bees, nest structure, natural enemies, predator, parasite

6-1.1 INTRODUCTION

Honey bees and stingless bees are classified in the family of Apidae. Their nests are often located in cavities. However, the structure and the composition of their nests may differ, even within a species. Honey bees (genus *Apis*) have three types of nesting: cavity-nesting with multi-combs (*Apis mellifera, A. koschevnikovi, A. nuluensis, A. indica,* and *A. cerana*), an exposed single comb nest on rocks or branches or tree trunks (*Apis dorsata, A. breviligua, A. binghami, A. laboriosa*) and a single comb nest on tree branches for dwarf honey bees (*A. florea* and *A. andreniformis*). Multi combs of cavity-nesting honeybees are built exclusively with wax (Oldroyd and Wongsiri 2006).

Stingless bees have at least two types of nests, i.e., cavity and base nests. A cavity nest is located on a trunk characterized by the outer entrance on trunk hollows. This type of nest is usually built at a certain height. A base nest is usually located below or at the base of the tree, characterized by an outer entrance attached to the outer wall of the tree and is usually hidden (Nunes *et al.* 2011). These bees also sometimes nest in termite and ants nests (Michener, 1974; Roubik 2006: Stingless bee nesting biology. Apidologie 37: 124-143.). They generally build their nests in tree holes (Buchwald and Breed 2005) and build their nest in wood, cracks in the walls, or under the residence roof (Kumar *et al.* 2012).

These bees have local names, e.g., klanceng (Central Java), lanceng (East Java), Teuweul (West Java), galo-galo (Sumatera), and kelulut (Kalimantan). Stingless bees are found in the tropics and subtropics (Free 1993). They are also widely distributed in Indonesia, i.e., 31 species in Kalimantan, 41 species in Sumatra, and nine species in Java (Schwarz 1937). Sakagami *et al.* (1990) reported six species of stingless bees found in Java, those are *Tetragonula laeviceps, Heterotrigona itama, Tetragonula drescheri, Tetrigona apicalis, Geniotrigona thoracica,* and *Lepidotrigona terminata.* Subgenera *Homotrigona, Lepidotrigona,* and *Heterotrigona* are restricted to tropical Asia (Michener 2000).

These bees can be found in the forests. Some types can also be adaptable in open forest areas, grasslands, and settlements (Inoue *et al.,* 1985). Some of these species also build nests underground (Jalil and Shuib 2014). The nest of stingless bees is constructed mainly with resin. The arrangement of the cells in the nest of stingless bees consists of cluster and comb form (Michener 1974). Nest is an artificial construction used by bees to lay eggs, spend time, and nurture [no nurturing occurs, they provide food, the queen lays an egg, and that is all] young bees. The nest of stingless bees consists of an entrance, cerumen, batumen, involucrum, storage pots, brood cells, and sometimes a scutellum and latrine (Roubik 2006). Cerumen is a mixture of propolis (resin) and wax to construct storage pots and brood cells. Batumen consists of thick resin and other materials such as soil or pebbles, as a layer of cavity nest (Roubik 2006, Jalil and Shuib 2014). Batumen and cerumen are used to protect the inner part of the hive. Involucrum is a layer around brood cells (Michener 2007).

Stingless bees are social insects that have an important role as plant pollinators (Wille 1983; Inoue et al. 1985) and have a high economic value related to producing honey and propolis. Honey of stingless bees has a distinctive scent with a lemon fragrance-like taste characterized by a mixture of sweet and sour flavour. Compared to the honey produced by honey bees, the honey of stingless bees often has a darker colour, more acidic taste and contains higher phenolic compounds (Garedew *et al.* (2003); Kek *et al.* (2014)). Propolis possesses anti-bacterial, antimicrobial and anti-fungal properties (Sforcin and Bankova, 2010). Stingless bees produce a higher amount of propolis than honey bees (Fatoni 2008). However, natural enemies like predators and parasites can reduce the productivity of bees, disturb the colony, affect the structure and composition of the nest or even kill the members of the colony.

Predators are animals that prey on other animals (Estes *et al.* 2001). Some predators attack the colony of stingless bees, i.e., wasps, ants, spiders, and centipedes. [sorry, but this is incomplete- big, bad mammals and other vertebrates are by far the most significant predators, followed by ants and perhaps termites]. Try consulting a book about this, like the tropical bee book by Roubik (1989). Centipedes may damage bee colonies within 2-5 seconds because they have a morphological variation and adaptation to take, catch, and hold prey (Kumar *et al.,* 2012). The bees usually do suicide bites as a defence mechanism against predators (Shackleton *et al.,* 2015). Parasites are organisms that live on other organisms. The parasites commonly found in honey bees are mites, fleas, moths, and small beetles (Strauss *et al.,* 2013). These parasites can be harmful to the health of honey bees (Shen *et al.*, 2005). Natural enemies like predators and parasites on the stingless bees have not been widely reported yet, so these researches are really important to conduct.

6-1.1.1 Aim of the Study

The research aimed to study the natural enemies and additional nest structures of stingless bees related to natural enemies.

6-1.2 MATERIALS AND METHODS

The stingless bee colony collection was made from August to November 2015. Ten colonies were taken from Pasuruan (East Java), eleven colonies from Tasikmalaya (West Java), and nineteen colonies from Rangkasbitung (Banten province).

6-1.2.2 Observation of Nest Structure of Stingless bees

Each colony of stingless bees was observed and classified based on their nest structure. Nothing to do with species or anything else. The structure and composition of the nest were observed, i.e., the outer entrance, inner entrance channel, storage pots (honey and pollen), and brood cells. Each colony observed then was moved to a wooden box, sized 30 cm x 25 cm x 15 cm. [which therefore totally disrupted all-natural architecture and other defences!]

6-1.2.3 Identification of Stingless Bees and Natural Enemies

Description of samples collected of stingless bees from each area were conducted based on the morphological study. Samples identification of stingless bees was conducted based on Sakagami (1990) and Michener (2007). Natural enemies were studied and collected from the stingless bees in bamboo. Afterwards, both samples, stingless bees and their natural enemies, were preserved in 70% ethanol. Samples were identified in the Laboratory of Entomology, Indonesian Institute of Sciences (LIPI) at Cibinong, Bogor. By whom? Is anyone remotely educated in stingless bee taxonomy?

6-1.2.4 Data Analysis

The stingless bees and natural enemies collected from each area were described. Nest structure and additional nest structure that arose by stingless bee natural enemies were also described.

6-1.3 RESULTS

A total of 39 colonies of stingless bees were identified as *Tetragonula* sp.; *Tetragonula* is a tiny stingless bee genus (about 3.5 mm body length) with a brown-blackish body and transparent wings. Thirty-nine colonies were found; 64% of the colonies have the common nest structure and the remaining (36%) with the additional nest structure (Table 21). Stingless bee colonies of Pasuruan and Tasikmalaya did not create the additional nest structure. On the contrary, more colonies of

Rangkasbitung created additional nest structures (73.68%), and the remaining had a common nest structure (26.32%) (Chart 12). Observation in the field showed that natural enemies were not found in the colonies of Pasuruan and Tasikmalaya. In contrast, more colonies of Rangkasbitung (52.63%) in bamboo were attacked by natural enemies (Table 22), compared to what in the others? Zero?

6-1.3.1 The Common Nests-Structure of Stingless Bees

Commonly, the nest structure of stingless bees in bamboo, from outer to inner, are nest entrance, cerumen, storage pots, brood cells, and cerumen (Figure 175). The nest entrance is a channel for in and out of the colony members. Storage pots consist of honey pots and pollen pots used by the colony as nutrient sources. Brood cells are smaller than storage pots used to lay eggs oviposited by the queen.

6-1.3.2 The Additional Nest Structures of Stingless Bees

The colonies of Rangkasbitung created the additional nest structure when their nest was damaged or attacked by natural enemies, such as predators and parasites. Results showed that some additional nest structures were inner entrance channels with much batumen and cerumen and created the root-like structure (Figure 177). What is the root-like structure for? An entrance tube? Take a good look at Roubik, D. W. 2021. Mutualism within parasitism within a mutualism: the bees and coccids that inhabit *Cecropia* ant-plants. *Ecology* 102(9):e03367. 10.1002/ecy.3367

6-1.3.3 Natural Enemies of Stingless Bees

Ten colonies in bamboos at Pasuruan and ten colonies at Tasikmalaya were found without their natural enemies. However, ten colonies at Rangkasbitung were found with their natural enemies, such as cockroaches (Blattodea: Blattidae), *Platysoma leconti* (Coleoptera: Histeridae), *Carpophilus* sp. (Coleoptera: Nitidulidae. We also found several eggs of lizard on the nest and larva of termites on below layers of the bamboo colony) (Table 22, Figure 176).

6-1.4 DISCUSSION

6-1.4.1 Nest Structure of Stingless Bees

Results showed that the nest structure and composition of stingless bees in bamboo, from the outside to the inside, consist of nest entrance, cerumen, storage pots (honey and pollen), brood cells, and storage pots. Generally, the structure and composition of stingless bee nests consist of nest entrance, inner tunnel, brood cells, storage pots (honey and pollen), and batumen layers (Sakagami *et al.* 1983).

The nests are generally built by materials, such as wax, resin, and mud. Wille (1983) stated that bees used olfactory cues to locate the resin and food sources (nectar and pollen). Lack of plant resins causes food storage pot and transparent brood cells (Leonhardt *et al.* 2010). Cerumen is a resin material mixed with wax. The wax of the stingless bee has a higher melting point than the wax of honey bees, *Apis mellifera* (Blomquist *et al.* 1985). If the wax is present in a higher concentration than resin, the texture of the nest becomes harder. The waxy nest can furthermore be hardened by mud to form batumen. While involucrum is cerumen are located around the brood cells to protect the nest from predators and parasites (Michener 2007).

There are two types of brood cells, cluster and comb form. Brood cells that do not contain eggs have darker colours, whereas those containing eggs are transparent. Chinh *et al.* (2005) stated that the new cells were yellow-brownish in colour but eroded the wax transparent. Cells are connected with another cell by pillars. Leonhardt *et al.* (2010) showed brood cells associated with one pillar with a small ball shape. Nectar contains carbohydrates adult bees use as energy sources to forage and defend their colony. In comparison, pollen contains protein used by young colony members (Jalil and Shuib 2014). The honey of stingless bees has more acidic and has a bright yellow (Chinh *et al.* 2005).

Additional nest structures were found on the bamboo colony in Rangkasbitung, Banten Province. Roubik (2006) reported that these bees build the architecture inside the hive to protect them from natural enemies. This structure is built in rotted cavities to protect the colony from predators and parasites. Chinh et al. (2005) reported that the inner surface is usually decorated with black cerumen thickness of 1.5-2 mm. In addition, the colonies used a lot of wax to avoid predators and parasites.

6-1.4.2 Natural Enemies of Stingless Bees

Natural enemies were not found in the colonies of Tasikmalaya (West Java) and Pasuruan (East Java). Contrary, some predators, such as cockroaches (Blattidae), histerid beetles (*Platysoma leconti*), and nitidulid beetles (*Carphophilus* sp.), were found in the colony of Rangkasbitung. Some lizard eggs were also found in the bamboo or box colonies. Ants (*Pheidole* sp.) were found in the stingless bee colonies that moved to a wooden box. Kumar et al. (2012) reported that some predators attacked stingless bee colonies, including wasps, ants, spiders, and centipedes in India. Ants are predators. If the number exceeds the limit, it can interfere with the colony of stingless bees.

The resin of plants is used as a defence mechanism against natural enemies (Roubik 2006; Leonhardt *et al.* 2010). If the colonies were disturbed by humans, workers showed defence behaviour, like biting and attacking the hair, eyes, nose, and ears (Wille 1983). The stingless bee workers also showed the defence mechanism against predators by standing and hovering. Standing is the behaviour of worker bees that keep the hive entrance. While hovering is the behaviour of worker bees that fly near the nest entrance. In the evening, some species of stingless bees close the outer entrance channel with cerumen[3] at a thickness of <0.05 mm (Grüter et al., 2010). [it isn't wax, and isn't <.05 mm]

6-1.4.3 Additional Nest Structures of Stingless Bees Arise as a Results of Natural Enemies

The shape of the nest entrance of stingless bees varied. These colonies build nest entrances surrounded by cerumen. Chinh *et al.* (2005) reported the shape of the nest entrance, and the internal nest entrance tube usually is not found in *Tetragonula laeviceps*. The nest entrance is usually found at the tree cavity an average height of 90 cm above the ground and or in artificial constructs of humans (Sakagami *et al.* 1983). No way this is typical or even relevant.

The nest structure of stingless bees has two main parts, i.e., storage pots and brood cells. Storage pots are usually found on the top and bottom of the nest, round form, thin walls, and dark brown. The thickness of storage pots is less than 1 mm. Brood cells are smaller than storage pots (Chinh *et al.*, 2005). Brood cells are usually open up and closed after the egg is laid. Commonly, brood cells are arranged regularly in the horizontal combs, and pollen pots are commonly found near the brood cells. Brood cells contain nutrients for the development of eggs and larvae, such as pollen, nectar, and honey (Michener 2000).

Additional nest structures are built to defend against predators. Several predators found in this study were cockroaches (Blattodea: Blattidae), histerid beetles *Platysoma leconti* (Coleoptera: Histeridae), nitidulid beetles *Carphophilus* sp. (Coleoptera: Nitidulidae) and ants *Pheidole* sp. (Hymenoptera: Formicidae). The beetle larvae live and grow in bees' colonies to obtain nutrient sources. At the same time, ants in the bee colonies aim to take a portion of food. Chinh *et al.* (2005) reported that worker bees defend their colony against ants by biting and using a resin to block the outer and inner entrance of the nest.

[3] Editor's note: brood cells unless from a very atypical species, even though white until pupal stage are made from cerumen, not wax.

6-1.5 CONCLUSIONS

The common nest structure of stingless bees in bamboo are nest entrance, cerumen, storage pots, brood cells, and storage pots. The nest of an underground colony of stingless bees is round-shaped, and storage pots and involucrum cover brood cells. The additional nest structure is formed in batumen, consists of inner entrance channels are lined by batumen or cerumen with a certain thickness, and a nest with a roots-like structure to trap the predators and parasites. Predators attacked the colonies of stingless bees in Rangkasbitung were cockroaches (*Blattidae*), histerid beetles (*Platysoma leconti*), nitidulid beetles (*Carpophilus* sp.), and ants (*Pheidole* sp.). You have some data, but your interpretation is ludicrous and full of academic and logical flaws. Forget it!

6-1.5.1 Acknowledgements

We thank the Biovillage Project of Indonesian Institute of Sciences, Indonesia endowment fund for education (LPDP) of Ministry of Finance, staff of Biology of Bogor Agricultural University (IPB) and staff of Zoology of Indonesian Institute of Sciences (LIPI) for permission and use of facilities during the research.

6-1.6 FIGURES AND TABLES

Table 21. Common nest structure and additional nest structure of stingless bee species from each sampling area.

Location of colony	Nest Structure	
	Common	Additional
Pasuruan	10	-
Tasikmalaya	10	-
Rangkasbitung	5	9
Total	25	14
Percentage (%)	64	36

Table 1. Common nest structure and additional nest structure of stingless bee species from each sampling area.

Table 22 A number of stingless bee colonies were attacked by natural enemies.

Natural enemies	the number of colonies attacked		
	Pasuruan	Tasikmalaya	Rangkasbitung
Number of colonies	10	10	19
Platysoma leconti (Coleoptera: Histeridae)	-	-	6
Carpophilus sp. (Coleoptera: Nitidulidae)	-	-	2
Cockroaches (Blattodea: Blattidea)	-	-	2
Total	-	-	10
Percentage (%)	-	-	52.63

Table 2 Some stingless bee colonies were attacked by natural enemies.

Figure 394 Structure and composition of the cavity-nesting of stingless bees in bamboo The nest entrance (a), storage pots (b), brood cells (c), and storage pots (d).

Structure and composition of the cavity-nesting of stingless bees in bamboo: The nest entrance (a), storage pots (b), brood cells (c), and storage pots (d).

Figure 395 Natural enemies attacked the stingless bee colonies in bamboo: Platysoma leconti (mummified) (a), P. leconti stuck sticky cerumen (b), eggs of lizard on the bamboo colony (c), eggs of lizard on the plastic cover of a wooden box (d), and larva of termites nesting on bamboo layers (e), and the cockroach (f).

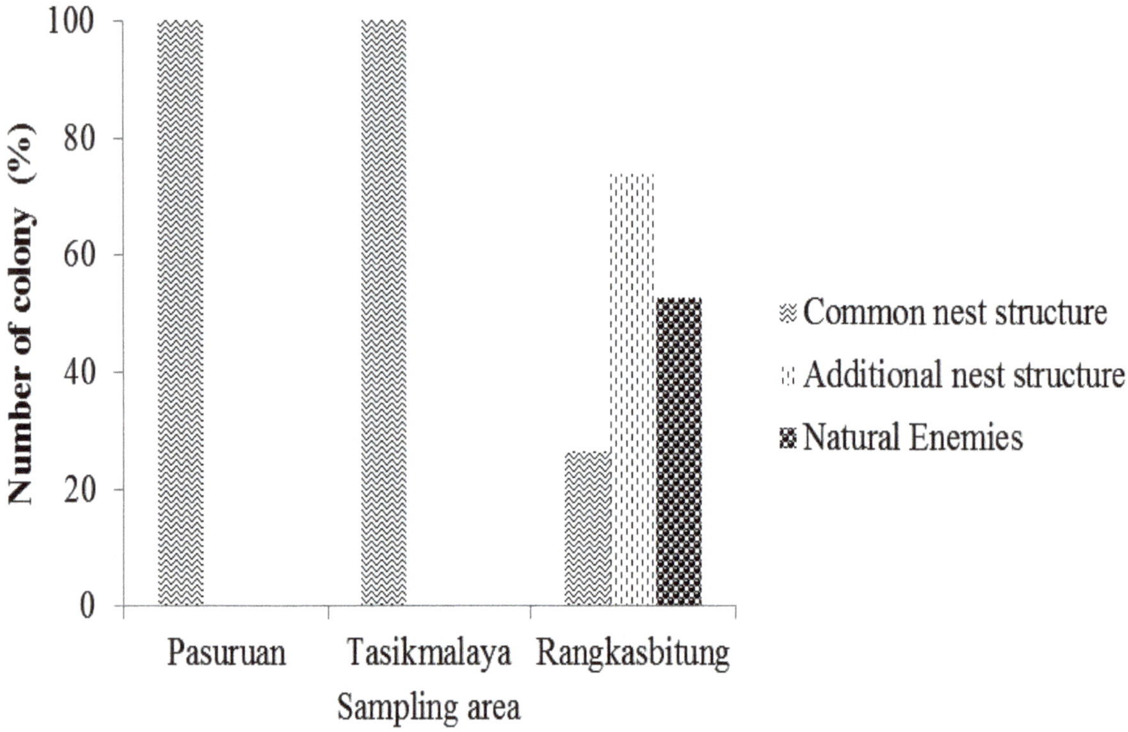

Chart 1 The relationship between the nest structure and natural enemies of stingless bee colonies.

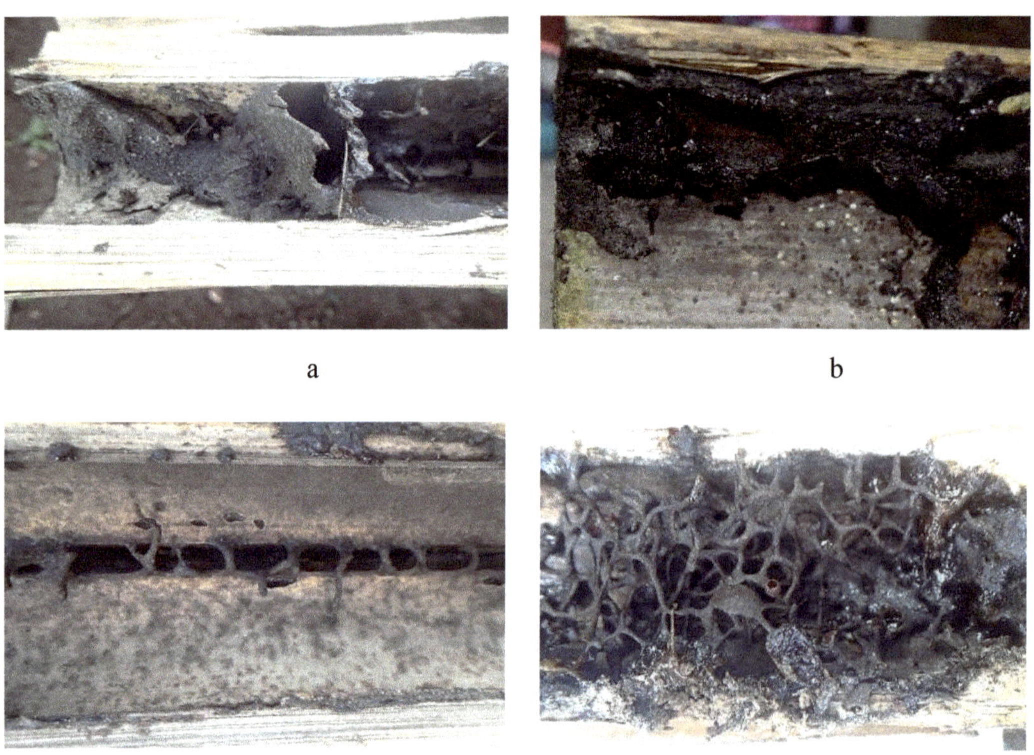

Figure 396 The additional nest structure of stingless bees: the inner entrance channel with batumen (a), the inner entrance channel with cerumen (b), nest with batumen (c), and root-like structure (d).

6-1.7 References

Blomquist GJ, Roubik DW, Buchmann SL. 1985. Wax Chemistry of Two Stingless bees of The *Trigonisca* Group (Apidae: Meliponinae). *Comp. Biochem. Physiol.* 82(1):137-142.

Buchwald R, Breed MD. 2005. Nestmate Recognition Cues in a Stingless Bee, *Trigona fulviventris. Anim. Behav.* 70:1331-1337.

Chinh TX, Sommeijer MJ, Boot WJ, Michener CD. 2005. Nest Architecture and Colony Characteristics of Three Stingless bees in North Vietnam with the First Description of the Nest of *Lisotrigona carpenteri* Engel (Hymenoptera: Apidae, Meliponini). *J. Kans. Entomol. Soc.* 3:26-39.

Estes J, Crooks K, Holt R. 2001. Predators, Ecological Role of. *Encyclopedia of Biodiversity.* 4:857-878.

Fatoni A. 2008. Pengaruh propolis *Trigona spp.* asal Bukittinggi terhadap beberapa bakteri usus halus sapi dan penelusuran komponen aktifnya [Tesis]. Bogor (ID): Program Pascasarjana, Institut Pertanian Bogor, Bogor.

Free JB. 1993. *Insect Pollination of Crops.* Cambridge (GB): Academic Press.

Garedew A, Schmolz E, Ingolf L. 2003. The Antimicrobial Activity of Honey of the Stingless Bee *Trigona* spp. *Journal of Apicultural Science* 47(1):37-49.

Grüter C, Karcher MH, Ratnieks FLW. Neotropical Entomology. 2010. The Natural History of Nest Defence in a Stingless Bee, *Tetragonisca angustula* (Latreille) (Hymenoptera: Apidae) with Two Distinct Types of Entrance Guards. 40(1):55-61.

Inoue T, Salmah S, Abbas I, Yusuf E. 1985. Foraging Behavior of Individual Workers and Foraging Dynamics of Colonies of Three Sumatran Stingless bees. *Res. Popul. Ecol.* 27:373-392.

Jalil A H, Shuib I. 2014. *Beescape for Meliponines*. Singapore [SG]: Partridge Publishing.

Kek SP, Chin NL, Yusof YA, Tan SW, Chua LS. 2014. Bees. Total Phenolic Contents and Colour Intensity of Malaysian Honey from the Apis spp. and Trigona spp. *Agriculture and Agricutural Science Procedia.* 2:150-155.

Kumar MS, Singh AJAR, Alagumuthu G. 2012. Traditional Beekeeping of Stingless Bee *(Trigona asp)* by Kani Tribes of Western Ghats, Tamil Nadu, India. *Indian Journal of Traditional Knowledge*. 11(2):342-345.

Leonhardt SD, Zeilhofer S, Blüthgen N, Schmitt T. 2010. Stingless Bees Use Terpenes as Olfactory Cues to Find Resin Sources. *Chem. Senses* 35: 603-611.

Michener CD. 1974. *The Social Behavior of the Bees*. Cambridge (GB): Harvard University Press

Michener CD. 2000. *The Bees of the World.* Edition 1. [Digital]. JHU Press.

Michener CD. 2007. *The Bees of the World.* Edition 2. Maryland (MD): JHU Press.

Nunes TM, Mateus S, Turatti IC, Morgan ED, Zucchi R. 2011. Nestmate Recognition in the Stingless Bee Frieseomelitta Varia (Hymenoptera, Apidae, Meliponini): Sources of Chemical Signals. *Animal Behaviour*. 81:463-467

Oldroyd BP, Wongsiri S. 2006. *Asian Honey Bees*. Cambridge (GB): Harvard University Press.

Roubik DW. Stingless bee nesting biology. 2006. *Apidologie*. 37:124-143.

Roubik DW. Stingless bee ecology (Apidae: Apinae; Meliponini). 2023. In press. *Ann. Rev. Entomol.*

Sakagami SF. 1983. Nest Architecture and Colony Composition of the Sumatran Stingless Bee Trigona (Tetragonula) laeviceps. *Kontyu Tokyo*. 51(1):100-114.

Sakagami SF, Ohgushi R, Roubik DW. 1990. *Natural History of Social Wasps and Bees in Equatorial Sumatra*. Sapporo (JP): Hokkaido University Press.

Schwarz HF. 1937. The Indo-Malayan Species of *Trigona*. *Bull Am Mus Nat Hist*. 76:83-141.

Sforcin JM, Bankova V. 2010. Propolis: Is There a Potential For the Development of New Drugs. *Journal of Ethnopharmacology*. 133:253-260.

Shackleton K, Toufailia HA, Balfour NJ, Nascimento FS, Alves DA, Francis L, Ratnieks W. 2015. Appetite for Self-Destruction: Suicidal Biting as a Nest Defense Strategy in Trigona *Stingless bees*. *Behav Ecol Sociobiol*. 69:273-281.

Shen M, Cui L, Ostiguy N, Cox-Foster D. 2005. Intricate Transmission Routes and Interactions Between Picorna-like Viruses (Kashmir Bee Virus and Sacbrood Virus) With the Honeybee Host and The Parasitic Varroa Mite. *Journal of General Virology*. 86:2281-2289.

Strauss U, Human H, Gauthier L, Crewe RM, Dretemann V, Park CWW. 2013. Seasonal Prevalence of Pathogens and Parasites in the Savannah Honeybee (*Apis mellifera scutellata*). Journal of Invertebrate Pathology. 114:45-52.

Wille A. 1983. Biology of The Stingless Bees. *Annual Review of Entomology*. 28:41-64.

Chapter 6-2

6-2.0 - Indonesian Propolis Content

Muhammad Yusuf Abduh[1,2], Fahmi Ramdhani[1]

[1]School of Life Sciences and Technology, Institut Teknologi Bandung, Jalan Ganesha 10, 40132 Bandung, Indonesia

[2]University Center of Excellence for Nutraceuticals, Bioscience and Biotechnology Research Center, Institut Teknologi Bandung, Jalan Ganesha 10, 40132 Bandung, Indonesia

According to Bankova et al. (2014), propolis does not have a specific chemical composition because the content of propolis depends on the vegetation in the area where bees are cultivated. In addition to local vegetation, different bee species and seasons also affect the chemical composition of propolis (Sulaeman et al., 2019). Tropical countries such as Indonesia have a high diversity of biological resources. Based on Kahono et al. (2018) research, as many as 46 species spread in Indonesia stingless bees (Meliponini). The large diversity of bee species and vegetation scattered in Indonesia makes propolis' content in Indonesia also different.

6-2.1 - Flavonoid and Phenolic Content of Indonesian Propolis

Several studies have examined the diversity of the content of stingless bee propolis in Indonesia. Abduh et al. (2021) have investigated the content in propolis from *T. laeviceps, T. biroi, T. drescheri* and *Pinus merkusii* resin which has been provided around the nests of the three species. Propolis extract from each species and pine resin used for bee forage was calculated for their flavonoid and phenolic content. Based on the research results of Abduh et al. (2021), different extraction methods produce different levels of flavonoids and phenolics. The phenolic and flavonoid content of propolis obtained by Abduh et al. (2021) is presented in table 1.

Table 1 *Levels of flavonoids and phenolics in propolis extracts of T. laeviceps, T. biroi, and T. drescheri*

Species	*Method*	*Phenolics (mg GAE/g)*	*Flavonoid (mg QE/g)*	*Reference*
T. laeviceps	*Maceration*	*343.93±44.32*	*35.77±9.94*	*Abduh et al. (2021)*
	Reflux	*270.97±119.3*	*30.58±5.53*	

	Soxhlet	136.48±13.90	20.12±0.95	
T. biroi	Maceration	123.81±24.69	5.48±3.31	Abduh et al. (2021)
	Reflux	98.13±48.82	4.41±0.55	
	Soxhlet	104.09±3.86	5.10±0.04	
T. drescheri	Maceration	225.87±70.36	14.57±9.30	Abduh et al. (2021)
	Reflux	170.22±70.34	12.11±3.83	
	Soxhlet	117.15±2.34	7.93±0.19	

Based on research by Abduh et al. (2021), a darker colour of the propolis extract indicates higher flavonoid content in the extract. This factor agrees with the study of Woo (2004), which states that a darker colour indicates higher flavonoid content.

According to Paviani et al. (2013), differences in the origin of propolis and differences in the polarity of the solvent used in the propolis extraction process affect the amount of yield and result in different types and amounts of flavonoids in analyzed content. Hasan et al. (2014), studied the flavonoid content of propolis from Makassar, Pekanbaru, Kendal, Pandeglang, and Banjarmasin. The flavonoid content, antioxidant activity, and anti-cytotoxicity activity of propolis originating from various regions in Indonesia are presented in table 2. Based on the research of Hasan et al. (2014), it can be concluded that the content of propolis in each region in Indonesia is different.

Tabel 2 *Total flavonoids, antioxidant activity, the anticytotoxic activity of propolis in various regions in Indonesia*

Characteristics	Makassar Propolis	Pekanbaru Propolis	Kendal Propolis	Pandeglang Propolis	Banjarmasin Propolis
Total Flavonoid (ug/ml)	38.78 ± 1.62	16.90 0.537	± 46.60 ± 0.78	30.62 1.50	24.60 ± 0.73
Antioxidant Activity (IC50) (ug/ml)	1125.56 ± 133	308.88 ± 12	144.06 52.53	± 68.935 5.63	4162.61 ± 845.9
Anticytotoxic activity at 100 g/ml (% live cells)	47.71 ± 9.31	76.35 ± 1.48	50.26 ± 2.70	70.64 ± 1.21	75.79 ± 1.33

6-2.2 - Chemical Compounds in Indonesian Propolis

Extracted propolis and *P. merkusii* resin provided as bee feed by Abduh et al. (2021) analyzed its chemical composition using GC-MS. Based on research by Abduh et al. (2021), it was concluded that there were similarities between the chemical compounds contained in the propolis extract and the *P. merkusii* resin provided by researchers around the cultivation site. The chemical compounds include alpha-pinene, camphene, 3-carene, alpha-terpinene, and alpha-terpineol, as presented in table 3.

Table 3. *Composition of pine resin and propolis extract produced by T. laeviceps, T. biroi, and T. drescheri*

Senyawa	*Propolis Extract*			Resin *Pinus merkusii*
	T. laeviceps	*T. biroi*	*T. drescheri*	
Alpha-Pinene	*1.60 – 2.22%*	*8.05 – 12.28%*	13.53 – 27.13%	*73.31%*
Camphene	-	-	*0.47%*	*1.20%*
Beta-Pinene	-	-	-	*0.89%*
3-Carene	-	*0.82 – 0.92*	*1.78 – 3.02%*	*12.94%*
Limonene	-	-	-	*1.10%*
Alpha-terpinene	-	-	*0.70%*	*1.27%*
4-Terpinenol	*0.20%*	*0.62%*	*0.62%*	
Alpha Terpineol	*0.39 – 0.68%*	*2.95 – 4.94%*	*4.79 – 5.59%*	*5.13%*
Octa-acid	-	*2.04%*	*6.19%*	-
decanoate		-	*10.12%*	
Abietic acid	*2.54%*	*0.23- 0,37%*	*0.29%*	-

Sulaeman et al. (2019) have analyzed the content of propolis originating from the provinces of South Sulawesi, South Kalimantan, and Banten. Propolis content was identified using the GC-MS tool. The compounds identified in samples from three regions of Indonesia include triterpenoids, n-alkanes, n-fat alcohols, n-alkanols, aromatics, n-aliphatics, methyl carboxylates, phenols, and ketoaldehydes. The content of compounds and the biological activity of propolis from the three provinces is presented in table 4.

Table 4 Identification of component activity in propolis extract samples with ethanol using GC-MS (Kalsum et al., 2016)

Compound	Compound Group	Propolis Extract			Bioactivity
		Sulawesi	Kalimantan	Banten	
Limonene	*Monoterpene Hydrocarbon*	*Yes*	-	Yes	*Immunomodulator, Anti-oxidant, Anti-bacterial, Anti-inflammation, Antiadenomic, Anti-Alzheimer, Antiasthma, Anti-cancer.*
Hexadecane	*Hydrocarbon Alkanes*	Yes			*Anti-Oxidant, Anti-Microbe*
Heptacosane	*Hydrocarbon Alkanes*	*Yes*			*Anti-bacterial*
Tetratetracontane	*Alkanes*		*Yes*	*Yes*	*Anti-oxidant, Hypoglycaemia*
1-Hexadecanol	*fatty Alcohol*	*Yes*	-	-	*Anti-oxidant*
1-Heptacosanol	*fatty Alcohol*	*Yes*	-	-	*Anti-oxidant, Antimicrobial, Anti-cancer*

Hexadecenoic Acid	*Aromatic Hydrocarbon*	-	*Yes*	*Yes*	Antioxidant, antiandrogen, taste buds, haemolytic inhibitor 5-alpha reductase, hypocholesterolaemia,
Acetic Acid	Aliphatic Acid	-	*Yes*	-	nematicides, pesticides, lubricants
Dioctyl Adipate	*Carboxylic Acid Esters*	*Yes*	-	-	*Antibacterial*
Nonacosane	*Aliphatic Hydrocarbons*	-	-	*Yes*	*Antimutagen*
Bis(2-ethylexyl) phthalate	*Aliphatic Hydrocarbons*	-	-	*Yes*	*Antimutagen, antileukemia*

(Contd.)

(Contd. Table 4)

Compound	Compound Group	Ekstrak propolis			Bioactivity
		Sulawesi	Kalimantan	Banten	
Propanoic acid 2-oxomethyl ester	*Carboxylate acid*	-	*Yes*	-	*Flavour*
2methyl Butanoic Acid	*Aromatic Hydrocarbon*	*Yes*	-	-	*Flavour*
Pentadecane	*Alkane hydrocarbon*	-	-	*Yes*	*Flavour*
1- (acetyloxy) -2-	*Alpha Ketoaldehyde*	-	*Yes*	-	*Flavour*
Propanone	*Phenols*	-	-	*Yes*	*Flavour*
2,6-dimethyl Phenol	*Phenolic Esters*	-	*Yes*	-	*Flavour*

Butanone

6-2.3 - Macronutrient Composition of Indonesian Propolis

Macronutrients can be used as ingredients in developing functional foods and beverages. Mulyati et al. (2021) studied the macronutrient composition of raw propolis from three regions in Indonesia, namely Riau Islands, Lampung, and South Sulawesi, with proximate analysis.

Table 5 presents the proximate composition of propolis studied by Mulyati et al. (2021).

Table 5 *The results of the proximate analysis of the three samples of propolis*

Parameter (%)	*Riau Islands Propolis*	*Lampung Propolis*	*S. Sulawesi Propolis*
Water/moisture	*8.76 ± 0.113*	*8.13 ± 0.013*	*7.44 ± 0.016*
ash	*1.31 ± 0.006*	*3.42 ± 0.016*	*2.07 ± 0.026*
protein	*2.04 ± 0.033*	*2.02 ± 0.041*	*3.05 ± 0.005*
fats	*38.67 ± 0.337*	*39.38 ± 0.13*	*61.64 ± 0.047*
fibre	*45.02 ± 0.653*	*57.66 ± 0.032*	*58.72 ± 0.472*
Carbohydrate	*49.21 ± 0.42*	*64.91 ± 0.49*	*25.80 ± 0.04*

Analysis of macronutrient content showed that Indonesian propolis generally contains high amounts of crude fat (38.67–61.64%), dietary fibre (45.02–58.72%), and carbohydrates (25.80–64.91%). The sample contains relatively low water (7.44-8.76%) and protein content (2.02-3.05%). In addition, the mineral quantity may be high because the ash content analysis shows more than 1%.

6-2.4 - Mineral Composition of Indonesian Propolis

Minerals are important micronutrients in human metabolism, especially as cofactors for enzyme activity. Measurement of mineral content was carried out to identify the potential of propolis as a mineral source. Information on the high mineral content in raw propolis forms the basis for its use in various applications. Developing propolis in powder or pill form can help meet the human body's needs. Indeed, each mineral has special benefits in metabolism due to its role as a cofactor of enzymes in our body (Berger et al., 2018).

Mulyati et al. (2021) tested the mineral content of propolis in the Riau Islands, Lampung, and South Sulawesi using Inductively Coupled Plasma – Optical Emission Spectrometry (ICP-OES) analysis. Table 6 shows a large variation in mineral content between propolis in Indonesia.

Table 6 *Mineral composition in Riau Islands propolis, Lampung, dan South Sulawesi*

Parameter (mg/100 kg)	*Riau Island Propolis*	*Lampung Propolis*	*S. Sulawesi Propolis*
Natrium (Na)	*73.915 ± 1.039*	*28.395 ± 0.318*	*15.53 ± 0.014*
Kalium (K)	*237.11 ± 3.309*	*-913.92 ± 3.734*	*118.23 ± 1.252*
Phosphorous (P)	*305.52 ± 0.11*	*1146.4 ± 13.371*	*146.4 ± 2.440*
Copper (Cu)	*13.725 ± 0.290*	*13.17 ± 0.042*	*12.875 ± 0.049*
Zinc (Zn)	*1.565 ± 0.035*	*99.26 ± 0.33*	*0.805 ± 0.007*
Iron (Fe)	*4.195 ± 0.120*	*8533.13 ± 76.16*	*3.89 ± 0.014*
Manganese (Mn)	*9.81 ± 0.240*	*23.685 ± 0.120*	*8.545 ± 0.078*
Calcium (Ca)	*173.1 ± 3.642*	*141.12 ± 0.31*	*99.335 ± 0.106*
Magnesium (Mg)	*74.36 ± 1.216*	*98.4 ± 0.636*	*39.065 ± 0.233*

The results of this study can be compared with previous research on the nutritional content of Indonesian propolis from East Java conducted by Halim et al. (2012). The sodium, potassium, copper, zinc, iron, and calcium content was significantly higher in propolis from the Riau Archipelago than the other samples ($p<0.05$). In addition, potassium, phosphorus, manganese and magnesium were very high in propolis from Lampung ($p<0.05$).

Halim et al. (2012) also studied Indonesian propolis bioactivity and nutritional content. The propolis studied came from the Wonosobo bee farm, East Java. Table 7 presents the heavy metal test results of Indonesian propolis extract compared with propolis from Brazil conducted by Halim et al. (2012).

Table 7 *Comparison of mineral content of Indonesian propolis extract with Brazilian propolis extract*

Parameter	*Unit*	*Indonesian Propolis*	*Brazilian Propolis*
Copper (Cu)	*mg/kg*	*4.51*	*1.07*
Zinc (Zn)	*mg/kg*	*35.2*	*1.61*
Manganese (Mn)	*mg/kg*	*5.98*	*0.30*
Iron (Fe)	*mg/kg*	*5.32*	*3.22*

Natrium (Na)	*mg/kg*	*56.6*	*34.0*
Calcium (Ca)	*mg/100 g*	*69.6*	*1.04*
Magnesium (mg)	*mg/100 g*	*80.6*	*1.86*

6-2.5 - Vitamin Composition of Indonesian Propolis

Vitamins are micronutrients needed in small amounts by the body to grow, develop and function normally. Vitamins are divided into two major groups, namely fat-soluble vitamins and water-soluble vitamins. Table 8 presents the vitamin content in Indonesian propolis extract compared with propolis from Brazil conducted by Halim et al. (2012).

Table 8 *Comparison of the mineral content of Indonesian propolis extract with Brazilian propolis extract*

Parameter	*Unit*	*Indonesian Propolis*	*Brazilian Propolis*
Vitamin A	*IU/100 g*	*<0.5*	*445*
Vitamin B1	*mg/100 g*	*1.10*	*<0.025*
Vitamin B2	*mg/100 g*	*2.89*	*<0.025*
Vitamin B6	*mg/100 g*	*0.35*	*<0.02*
Vitamin C	*mg/100 g*	*5.51*	*<0.2*
Vitamin E	*mg/kg*	*4.09*	*0.40*

B-complex vitamins are vitamins needed by our bodies that are soluble in water. Mulyati et al. (2021) have studied the content of vitamin B complex in propolis in Indonesia. Table 9 presents the B-complex vitamin profile of propolis as studied by Mulyati et al. (2021).

Table 9 *Profile of vitamin B complex from propolis Riau Islands, Lampung, dan South Sulawesi*

Vitamin	*Riau Islands Propolis*	*Lampung Propolis*	*South Sulawesi Propoli*
B1 (Thiamin)	-	-	-
B2 (Riboflavin)	-	-	-
B3 (Nicotinamide) *(mg/100 g)*	*7.99 ± 0.11*	*5.88 ± 0.1*	*14.29* *0.36*

B3 (Niacin) (mg/100 g)	*2.71 ± 0.08*	*-*	*12.10 0.31*
B6 (Pyridoxine) (mg/100 g)	*2.32 0.01*	*99.26 ± 0.33*	*5.67 ± 0.04*
B9 (Folic acid) (ug/100 g)	*-*	*8533.13 76.16*	*-*

The research results by Mulyati et al. (2021) showed that thiamine and riboflavin were not detected in all propolis samples. However, nicotinamide and pyridoxine were found in all propolis samples. The pyridoxine content in propolis from Lampung was detected significantly (99.25 mg/100 g) ($p<0.05$). This concentration shows that consuming 1.31 g of raw propolis from Lampung can meet 100% of the RDA (RDA for pyridoxine is 1.3 mg). In addition, propolis from Lampung contains higher folic acid (8533.13 g/100 g) compared to other samples ($p<0.05$). Folic acid has an important role in reducing the risk of congenital disabilities, cardiovascular disease, and cancer (Kennedy, 2016). In addition, the niacin content was only found in propolis from the Riau Islands and South Sulawesi. The niacin content in propolis from South Sulawesi was significantly higher than the other samples ($p<0.05$).

[Too bad it cannot be digested by humans, or can it, and how? Terpenoids make you sick and challenge the immune system, is that all?]

6-2.6 References for Propolis content

Abduh, M. Y., Zainuddin, I. M., Ramdhani, F., Setiawan, A., & Rifqialdi, G. (2021). Extraction of Propolis Produced by *Tetragonula* spp. Cultivated in Modular *Tetragonula* Hives. 1st International Conference on Bioscience and Biorefinery 2021. Bandung.

Bankova, V. (2009). The chemical diversity of propolis makes it a valuable source of new biologically active compounds. Journal of Api Product and Api Medical Science, 1(2), 23-28.

Berger, M. M., Sousse, L. E., Klein, G. L., Branski, L. K., & Herndon, D. N. (2018). Micronutrient Homeostasis. Total Burn Care, 276-279.

Halim, E., Hardinsyah, H., Sutandyo, N., Sulaeman, A., Artika, M., & Harahap, Y. (2012). Kajian bioaktif dan zat gizi propolis Indonesia dan Brasil. Jurnal Gizi dan Pangan, 7(1), 1-7.

Hasan, A. E., Mangunwidjaja, D., Sunarti, T. C., Suparno, O., & Setiyono, A. (2014). Investigating the antioxidant and anticytotoxic activities of propolis collected from five regions of Indonesia and their abilities to induce apoptosis. Journal of Food and Agriculture, 26(5), 390-398.

Kahono, S., Chantawannakul, P., & Engel, M. S. (2018). Social Bees and the Current Status of Beekeeping in Indonesia. In P. Chantawannakul, G. Williams, & P. Neumann, Asian Beekeeping in the 21st Century (pp. 287-306). Berlin: Springer Verlag.

Kalsum, N., Sulaeman, A., Setiawan, B., & Wibawan, I. W. (2016). Phytochemical profiles of propolis *Trigona* spp. from three regions in Indonesia using GC-MS. J Biol Agric Health, 6(14), 31-37.

Kennedy, D. O. (2016). B vitamins and the brain: mechanisms, dose and efficacy—a review. Nutrients, 8(2), 68.

Mulyati, A. H., Sulaeman, A., Marliyati, S. A., Rafi, M., & Rokhmah, U. F. (2021). Macro and Micronutrient Content of Raw Propolis Collected from Different Regions in Indonesia. Journal Gizi Pangan, 109-114.

Paviani, L., Fiorito, G., Sacoda, P., & Cabral, F. (2013). Different solvents for extraction of Brazilian green propolis: composition and extraction yield of phenolic compounds. In III IberoamericanConference on Supercritical Fluid, 1-5.

Sulaeman, A., Fikri, A. M., Kalsum, N., & Mahani, M. (2018). *Trigona* Propolis and its Potency For Health And Healing Process. In R. B. Singh, R. R. Watson, & T. Takahashi, The Role of Functional Food Security in Global Health (pp. 425-448). Academic Press.

Woo, K. S. (2004). Use of bee venom and propolis for apitherapy in Korea. Asian Apicultural Association Conference and BEENET Symposium and Technofora, College, Laguna (Philippines (pp. 311-315). Seoul: UPLB-BP.

CHAPTER 6-3

6-3.0 - Propolis as Antibacterial and Anti Breast Cancer Substance

Akhmad Endang Zainal Hasan[1]*, Amin Fatoni2, Gerardus Diri Tukan[3], Syaiful Abidin1, Rizki Prasetyo[1], Laksmi Ambarsari[1], I Made Artika[1], Fitriannur[1], Kuswandi[4], Djumali Mangunwidjaja[5], Titi Candra Sunarti[5], Ono Suparno[5], Agus Setiyono[6], Haryanto Susilo[7], Adinda Virginia Dwi Setyo[7], Henry Adprabowo[1], Metty Lasmiyanti[1], Poppi A.K.[1], Kasno[1], Adji Suranto[8]; Dian Anggraeni[1], Henry Adiprabowo[1], Endang[1], Ade Jamaludin Sunny[1]

[1] Fakultas Matematika dan Ilmu Pengetahuan alam, Institut Pertanian Bogor

[2] Fakultas MIPA, Universitas Soedirman, Purwokerto

[3] Universitas Katolik Widya Mandira (UNWIRA), Kupang

[4] Balai Penelitian Peternakan, Bogor

[5] Fakultas Teknologi Pertanian, Institut Pertanian Bogor

[6] Fakultas Kedokteran Hewan, Institut Pertanian Bogor

[7] Fakultas MIPA, Universitas Pakuan, Bogor

[8] Rumah Sakit Cipto Mangoen Koesoemo, Jakarta

*Corresponding author: pakzainalhasan@gmail.com

6-3.1 Propolis

Bees are animals that can produce various natural products, one of which is propolis. Propolis is a substance that contains wax and resin, is collected from plant sources, and is sticky. The word Propolis comes from the Greek word pro, which means in front and polis, which means city. This name cannot be separated from the usefulness of propolis as a protective substance in front of honey bee hives, both against unfavourable environmental threats and attacks by other organisms. Another function of propolis is to support the growth of honey bees, blanket a carcass so that it does not cause disease, and protect against sunlight. The function of propolis as a protector makes propolis appropriately placed at the entrances of the hive and its edges (Suranto 2010; Siregar et al. 2011).

Bees carry propolis to the hive in their pollen baskets, but the source of that propolis is often uncertain. Three well-known theories explain the source of Propolis, namely Kustenmacher's theory, Rosch's theory, and Phillip's theory. Kustenmcher theory says that propolis resin comes from pollen granules, while Rosch's theory explains that propolis resin comes from trees. Phillip's theory combines two: the resin comes from pollen and trees (Suranto 2010; Sarwono 2001). All miss the point entirely. The CHEMICALS CONTAINED in propolis have a plant origin, and also a bee origin, and it is in resin, gums and pollen and also nectar which, in the stingless bees, can be contained in the storage pots (and brood cells) of the nest. Apis has very little capacity here. Within the nest, the chemicals are exchanged between cerumen and honey and pollen stored in pots, and the recycled material from cerumen then becomes part of the nest environment. Period. Not many people appear to understand this basic fact.

Propolis contains several organic compounds, including resins, waxes and fatty acids, essential oils, pollen, and other compounds. The resin in propolis (as much as 45 to 55%) consists of several types of flavonoid compounds, phenolic acids, and esters. The type of compound in the essential oil of propolis is volatile (as much as 10%). Pollen contains 5% protein and 1% amino acids.[NO, up to 30-40% protein, and 20% aas] Other compounds in propolis (as much as 5%) include minerals, Au, Ag, Cs, Hg, La, Sb, ketones, lactones, quinones, steroids, benzoic acid, esters, vitamins, and sugars (Trubus 2010). [which are not FREE but tied up with the matrix of terpenes]

The content possessed by Propolis makes Propolis widely used by humans. Propolis protects the skin from all kinds of free radicals, restore collagen function, and eliminate acne, black spots, and wrinkles (Hasan 2010). [and if it is eaten, you are not happy]Propolis also treats AIDS, diabetes mellitus, pneumonia, herpes, bronchitis, haemorrhoids, coronary heart disease, and cancer. [this is still not well known or at all substantiated, by proper tests and isolation of active ingredients]. This is propaganda, still. Another benefit of propolis is a furniture preservative because of its antifungal properties. Other benefits of propolis are gold and silver polish, larvae traps, and baking ingredients (Suranto 2010).

6-3.2 Flavonoids

Flavonoids are the most abundant phenolic compounds found in nature. About 2% of the total carbon synthesized by plants (about 1 x 10 tons/year) is converted into flavonoids or related compounds. Flavonoids contain 15 carbon atoms in their basic nucleus, with about 1 x 10 C6-C3-C6 arrangement. The hydroxyl group is always present at carbon number 5 and number 7 in ring A, while in ring B, the hydroxyl or alkoxyl group is found at carbon number 3 and number 4 (Sirait 2007; Markham 2006).

Flavonoids are classified into 11 variants: flavones, flavonols, flavanones, flavanonols, isoflavones, chalcones, dihydrochalcones, and have anthocyanidins, catechins, and flavan-3,4-diols. All these variants are interrelated because of the same biosynthetic pathway, namely the shikimate pathway and the acetate-malonate pathway. Flavonoids were synthesized from chalcone precursors with the help of chalcone synthetase enzymes from p-coumaryl-CoA and three CO2 molecules. Chalcone is converted to flavonone with the help of chalcone isomerase. The ring structure is formed by adding a phenolic hydroxyl group to the double bond of the carbon ring connecting the two phenolic rings. Flavanones become precursors of other flavonoid variants through several stages (Helt and Piechulla 2010).

Propolis contains the most flavonoids compared to other bee products such as honey, royal jelly, and pollen (Siregar et al., 2011). Flavonoids in propolis ranged from 10 to 20% higher than pollen (5%)

and honey (0.006%) (Suranto 2010). The flavonoids contained in propolis include flavones (crisin, apigenin, luteolin, asacetin, and ermanin), flavonols (rutin, morin, quercetin, myricetin, kaempferol, galanin, and pinobaksin), flavanones (naringin, naringenin, hesperetin, pinocembrin, and pinosrobin) and isoflavones (daidzein and genistein) (Chang et al. 2002; Hasan 2013).

The flavonoids contained in propolis have various benefits, namely fortifying the body from germ attacks, preventing the formation of prostaglandins that cause pain and stimulating white blood cells. Another function of these flavonoids is to prevent allergies by preventing the leakage of histamine and serotonin. Bioflavonoids in propolis are also useful in reducing inflammation of the gums, canker sores, and oral disorders. Other benefits of propolis bioflavonoids are pain relief, anti-inflammatory, anti-allergic, antacid, antibiotic, antifungal, anti-radiation, antioxidant, antiseptic, antitumor, anticancer, and immunostimulant (Suranto 2010).

6-3.3 Functional Propolis

The role of propolis as an antibacterial can inhibit the growth of *E. sakazakii*. The nature and characteristics of *E. sakazakii* have been extensively studied for its physiological properties and resistance in the environment. The study results showed that propolis was extracted from meliponine beehives by maceration using 70% alcohol (propolis ethanol extract) and yielded a 17.76% yield. Antibacterial activity test was carried out by agar well diffusion method on three bacterial isolates of *E. sakazakii* IB-19b, IB-29a, and ATCC 35217. Through preliminary tests, propolis activity was higher than commercial Propolis and lower than ampicillin. The effectiveness of propolis against commercial propolis on isolates IB-19b, IB-29a, and ATCC 35217 were 143.99%, 169.65%, and 155.76%, respectively. The effectiveness of propolis extract against ampicillin in isolates IB-19b, IB-29a, and ATCC 35217 were 39.37%, 38.94%, and 47.37%, respectively. The minimum growth inhibition concentration obtained in each isolate was different. The minimum growth inhibition concentration for isolates IB-19b, IB-29a, and ATCC 35217 were 12.50%, 25.00%, and 50.00%, respectively.

The next research is to test the ethanol extract of meliponine propolis (EEP) from Bukittinggi, West Sumatra, against *Campylobacter* spp. The antibacterial activity of EEP was measured by the disc diffusion method. The group of propolis compounds was also analyzed for the presence of alkaloids, flavonoids, saponins, tannins, steroids, and terpenoids. This study revealed that meliponine EEP showed antibacterial activity against *Campylobacter* spp. The compound groups detected in the EEP were flavonoids and tannins. The indication that meliponine propolis has antibacterial activity is probably because of this compound.

Another role of propolis as an antibacterial has shown antibacterial activity on *Salmonella* sp. with a minimum inhibitory concentration of 0.87% (w/w). Propolis is a flavonoid and tannin that shows antibacterial activity. The group of compounds was detected in the ethanol extract. The role of propolis also functions in inhibiting the periodontal disease, including dental caries, which is still a major problem in the world. Usually, commercial toothpaste contains fluoride, which plays an important role in preventing tooth decay. But an overdose of fluoride causes fluorosis, bone breakdown and anaemia. The results showed that based on phytochemical analysis, the ethanolic extract of propolis (EEP) contained various bioactive compounds such as flavonoids, tannins, alkaloids, triterpenoids and saponins. Extract effectiveness was 96.5% relative to propolis X, 41.04% relative to 10 mg/ml ampicillin and 240.57% relative to 3000 ppm NaF. The Minimum Inhibitory Concentration of extract was 3.13%. Therefore, propolis extract can be used as an anti-caries additive in toothpaste. Statistical analysis showed that 100% EEP, X propolis and ampicillin 10 mg/ml could reduce *S. mutans* colonies.

Studies on lactic acid and pathogenic bacteria have different responses to propolis levels. Although the activity of propolis depends on its concentration, the lower the concentration, the less the antibacterial activity. However, several studies have shown something that surprises us. The results from the consumption of small amounts of propolis do not need to be doubted. Propolis at a concentration of 0.6% stimulated the growth of *Lactobacillus casei* subsp. *rhamnosus* (23.725x108 cells/mL), but inhibited the production of lactic acid, on average (0.071%) lower than the control (0.149%). Propolis did not affect the growth of *Streptococcus thermophillus*, but propolis concentration of 0.6% stimulated the production of lactic acid on average (0.182%), higher than the control (0.112%). Propolis inhibits the growth of *Lactobacillus delbrueckii* subsp. *bulgaricus*, but at a concentration of 0.2%, the total growth was the highest (3.775x108 cells/mL), and the average stimulated lactic acid production (0.195%) was higher than the control (0.123%). It is appropriate for propolis to be made as candy that can be consumed in small quantities but is still functional.

The propolis antibacterial role was tested against *Streptococcus mutans*, which is thought to play an important role in the aetiology of dental plaque development in humans associated with dental caries, the most common oral disease. The extract was then coated and homogenized to form propolis microparticles. These particles were then mixed into chewy candy preparations in chewy propolis candy. The antibacterial activity of the candy was further tested in vitro against *S. mutans* isolated from dental plaque. The results showed that the ethanolic extract of propolis from the beehive of stingless bees could be homogenized to form propolis microparticles. Propolis microparticles can be used as a supplement in the formulation of chewy propolis candy. The chewing gum with propolis showed antibacterial activity against *S. mutans.*

Another study aimed to determine the potential of meliponine nanopropolis from Pandeglang as an antibacterial. A high-speed homogenization technique makes nanopropolis with maltodextrin coating and solvent evaporation technique. An antibacterial activity test was carried out by the agar well diffusion method. SEM results showed nanopropolis measuring 175-873 nm. The results obtained that nanopropolis acts as an antibacterial, both Gram-positive and Gram-negative. Nanopropolis has higher effectiveness than propolis extract by 208.86%, 211.83%, 227.01%, and 230.29% against *B. subtilis*, *S. aureus*, *E. coli*, and *Salmonella* sp. The effectiveness of nanopropolis, when compared with ampicillin 10 mg/ml, was 43.87%, 49.12%, 42.35%, and 37.58%. Minimum Growth Inhibitory Concentrations (KHTM) were 0.15%, 0.075%, 0.15%, and 0.313% against *B. subtilis, S. aureus, E. coli,* and *Salmonella* sp.

The role of propolis that can be revealed is that propolis can inhibit cancer cells. However, the geographical origin of propolis determines its biological properties and the content of its active ingredients. In a study involving propolis collected from five regions of Indonesia. Propolis from Pekanbaru has a higher yield than other regions with a value of 19.97%; propolis from Kendal has a higher quantity with a value of 46.60%, total flavonoid content; propolis from Pandeglang was higher in DPPH oxidation capacity with a value of 68.94 g.ml-1; propolis from Kendal, showed the induction of petite cells in S. cerevisiae cells with a value of 81.44%, and the best anticytotoxicity against MCF-7 breast cancer cells was observed in propolis from the Makassar region with a live cell value of 47.71%. All propolis extracted from stingless beehives from five regions in Indonesia contains flavonoids.

The research on the role of propolis in inhibiting cancer cells was studied and used the RSM method to determine the extraction parameters that could produce the optimum propolis ethanol extract. The optimum extraction time and a 70% honeycomb ethanol solvent ratio were selected to induce apoptosis for 27 minutes and 20, respectively, with the yield of propolis extract as much as 12.67%. The ability of propolis extract to inhibit as much as 50% (IC50) or kill 50% of Michigan Cancer Foundation-7 (MCF-7) cancer cells and induce Apoptosis of S. cerevisiae each was 75.34 g. mL-1, 233 g. mL-1, and 6,015 g.mL-1. The results of the in-vivo efficacy test against cancer cells induced by 7,12-dimethyl-benz(a)anthracene (DMBA) showed that the propolis concentration of 233 g.mL-1 could heal tissue damaged by tumours. The research was then continued in vivo using experimental mice and nanoparticle-sized propolis and propolis not formed by nanoparticles. The result was that the effect of nanopropolis at a dose of 32 g/mL and a dose of propolis 233 g/mL was similar in reducing tumour size, healing tumour wounds and eliminating cancer cells. It turns out that there is a relationship between the particle size of the absorbent material. The results showed that nano propolis with a small concentration effectively treats mammary gland tumours and breast cancer.

6-3.4 References

Chang CC, Yang MT, Wen HM, Chern JC. 2002. Estimation of total flavonoid content in propolis by two complementary colourimetric methods. *J Food Drug Anal*. 10(3): 178-182.

Fatoni, A., I M. Artika, A.E.Z. Hasan, Kuswandi. 2008. Antibacterial Activity of Propolis Produced by MeliponineAgainst Campylobacter spp. HAYATI Journal of Biosciences, 15 (4);161-164

Hasan A.E.Z., Artika I.M., Fatoni A., Kuswandi, Haryanto B. 2011. Antibacterial Activity of Propolis *Meliponine*from Bukittingi West Sumatera Against *Salmonella* sp. *Chem Prog*, 4(1): 55-59.

Hasan A.E.Z., Mangunwidjaja D., Sunarti T.C., Suparno O., Setiyono A. 2016. Antibreast Cancer Activity of Nanopropolis Indonesia on induced Mammary Gland Tumor by DMBA in virgin *Sprague-Dawley* rats. *Biotropia*, 23(1): 35-41.

Hasan A.E.Z., Mangunwidjaja D., Sunarti T.C., Suparno O., Sutiyono A. 2014. Investigating The Antioxidant and Anticytotoxic Activities of Propolis Collected from Five Region of Indonesia and Their Abilities to Induce Apoptosis. *Emir J Food Agric*, 26(5): 390-398.

Hasan AEZ. 2013. Pembuatan Nanopropolis Asal Indonesia sebagai Bahan Antikanker Payudara. [Disertasi]. Bogor (ID): Institut Pertanian Bogor

Hasan. A.E.Z., Mangunwidjaja D., Sunarti. T.C., Suparno O., Setiyono A. 2013. Optimasi Ekstraksi Propolis Menggunakan Cara Maserasi Dengan Pelarut Etanol 70% Dan Pemanasan Gelombang Mikro Serta Karakterisasinya Sebagai Bahan Antikanker Payudara. *J Tek Ind Pert.* 23 (1):13-21.

Hasan. A.E.Z., Mangunwidjaja D., Sunarti. T.C., Suparno O., Setiyono A. 2013. Pembuatan Nanopropolis Asal Indonesia Sebagai Bahan Antikanker Payudara. *Disertasi*. Bogor: Sekolah Pascasarjana Institut Pertanian Bogor.

Helt HW, Piechulla B. 2010. *Plant Biochemistry* 4th Edn. Massachusetts (US): Academic Press.

Markham RK, Andersen OM. 2006. *Flavonoids: Chemistry Biochemistry Applications.* Florida (US): CRC.

Sirait M. 2007. *Penuntun Fitokimia dalam Farmasi*. Bandung (ID): ITB Press.

Siregar HCH, Fuah AM, Octaviany Y. 2011. *Propolis: Madu Multikhasiat*. Jakarta (ID): Penebar Swadaya.

Suranto A. 2010. Dahsyatnya Propolis untuk Menggempur Penyakit.Jakarta (ID): Agromedia Pustaka.

Trubus. 2010. Propolis dari Lebah tanpa Sengat, Cara Ternak dan Olah. Jakarta (ID): Trubus.

Trusheva, B., D. Trunkova, V. Bankova. 2007. Different extraction methods of biologically active components from Propolis: a preliminary study. *Chem Central J.* 1(13): 1-4.

Chapter 6-4

6-4.0 THE TROPICAL MELIPONICULTURE ECO-SYSTEM BUBBLE: Overcoming Vulnerabilities and Threats

By Abu Hassan Jalil

6-4.1 - INTRODUCTION

Meliponiculture involves multiple activities that complicate the activity and require a systematic plan to be understood and implemented either as a hobby or for commercial purposes. To simplify all activities, which involve vulnerabilities and threats (countermeasures and retardants), and to illustrate all the output and subsequent impact, a schematic diagram has been designed to represent in one figure as shown below (**Figure 17.1**). It is known as the 'Tropical Meliponiculture Eco-System Bubble'.

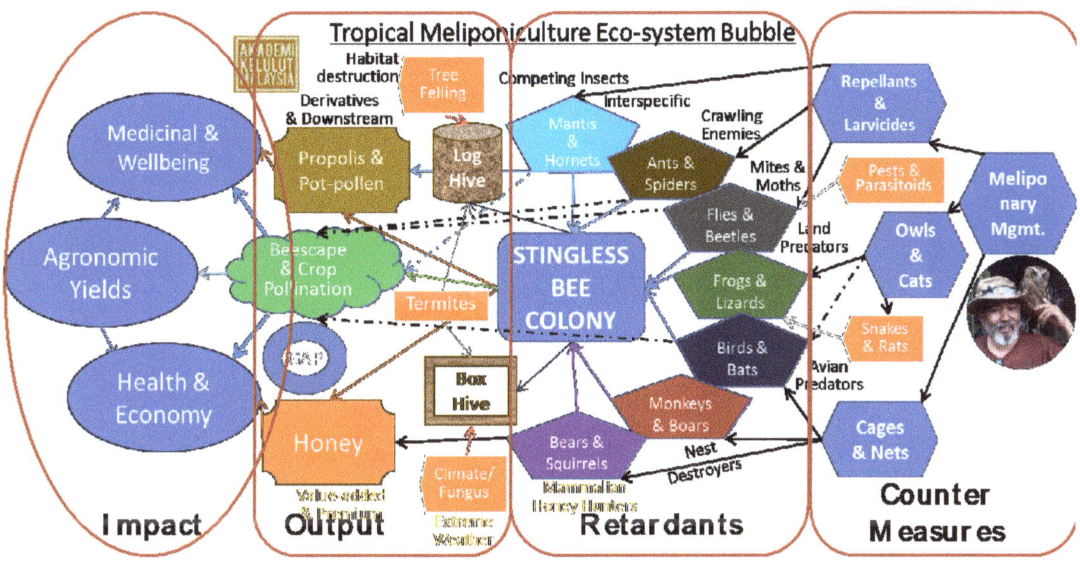

Figure 397 Tropical Meliponiculture Eco-System Bubble

The figure has three main components: the retardants, the output, and the impact. The predators' and enemies' food chain with each colour legend is at the right. The light blue hexes are beneficiaries, the derivatives are golden shields, and the green cloud is related to vegetation forage. The final light blue ovals are a positive impact on the left! The orange boxes with white borders are indirect cause and effects.

Each predator/enemy category overlap with the adjacent,

- The bears interact with monkeys
- Monkeys interact with birds and bats
- Birds can be predators to small frogs
- Frogs and lizards take up flies n beetles

- Flies and beetles are trapped and consumed by spiders
- Ants compete with other insects in almost every activity

The pentagon is an arrowhead and overlaps the corner in the predators' part over the next. Anyways, a natural ecosystem is complex. God has tremendous wisdom. We humans need to develop a simple model to explain human domestication.

6-4. 2 - CONCEPTUAL BACKGROUND

By understanding the Tropical Meliponiculture Eco-System Bubble, a few aspects in the meliponiculture should be considered as follows:

Initial setbacks & Retardants:

- Habitat destruction, Deforestation, Tree felling
- Intraspecific competition, Overcrowding & Aggressive swarming
- Pesticides, Hive shifting, Entrance closing

Pollination:

- Two categories; Natural pollination & Forest conservation
- Planned pollination in selected crops. Pollination under natural conditions also leads to the sustenance of plant biodiversity.

The general theme is the meliponiculture ecosystem as a human intervention and domestication activity. Besides the abovementioned aspects, meliponiculture activity will involve interspecific competition, intraspecific competition, pesticides and pollination peculiarities, and the others in meliponary management and GAP - Good Agriculture Practice. It is quite tough to include natural ecology, which makes the bubble more complicated (it might burst), or the elements too small for detection. In the meantime, the categories depicted are representative.

Heat, the cold result of climate extremes and colony placement, plus noise and smoke, will come under Meliponary Management. Other types of bees are in the 'competing insects' category, such as Apidae tribes like Apinini, Ceratinini, Bombini, and Vespidae are major foraging competitors (Vit, Roubik, & Pedro-(eds), 2013). Carpenter bees (Xylocopinae) and leafcutter bees (Megachilidae) sometimes rob stingless bee propolis and compete for nectar forage. Intraspecies compatibility is in Meliponary Management (Roubik, 2006) (Thai, 2001).

Proper landscaping practices are needed to safeguard, acclimatize and conserve these bees, especially the endemic stingless bee species. We cannot tell predators and insectivorous birds or mammals not to devour the bees, but we can tell the bees in unspoken communication through planting and organization. As an example, consider that birds usually capture bees above the canopy of trees.

We can tell our stingless bees not to go where the birds frequent. We do this by planting downward blooming flowers like Syzygiums, low flowering shrubs and beneficial ground covers like Portulacas and Zephyrantes (Jalil, 2014). We devise all sorts of traps and deterrents at their nests and hives to prevent and reduce predatory activity.

From bird trap nets to phorid fly traps at the beehive stations, their foraging of apical flowers is much desired in bird predation. It may not be so difficult in the case of palms and coconut trees. Palm tree flowers are favourites of stingless bees and are always under the protection of the palm canopy. The height of these palm blooms is a great deterrent to frogs and toads. Cleverly designed barriers (attached to the palm trunks) can prevent climbing frogs and lizards or chameleons from ambushing the foraging bees.

A well-designed floral landscape can also help the bees conserve their energy. Thus, the bees' efforts can go more toward regulating hive temperatures, nest building and brood health. Avoiding predating birds and freeing themselves from spider webs are stressful (or fatal) in the life of a stingless bee. Reducing this stress allows them to pool their efforts into building stronger colonies. Assuming that the bee colony as a single superorganism will require help in the artificial environment in which it has been placed, it can manage survival in the three-dimensional zone of varied and endless predators and dangers which although reduced in that artificial environment, still exist.

A few steps will be explained below to successfully observe the Tropical Meliponiculture Eco-System Bubble, starting with initializations, primary output, the secondary output, threats, and the meliponiculture management and industry impact.

6-4. 3 - INITIALIZATION

6-4.3.1 - Bee Housing

The stingless bee colony must be initialized in an appropriate hive to start a meliponiculture concern (Michener, 1961). Depending on the beekeeper's ultimate goal, he may keep in a log hive and have a topping box to harvest products from the hive or culture the colony in a portable box hive to move to sites for pollination services (Jalil, & Roubik(ed), 2016, 2018, 2019). These are the host and housing components of the core meliponiculture initiation (Roubik, 2006).

6-4.3 - PRIMARY OUTPUT

6-4.3.1 - Propolis and Pot-pollen

Propolis and pot-pollen are direct outputs resulting from interactions between stingless bees with 'Beescape' vegetation. It has both medicinal and commercial value.

6-4.3.2 - Honey

Honey is the key motivating output in meliponiculture and is a by-product of the pollination activity. Its commercial value is a sought-after commodity in the meliponiculture industry (Vit, Roubik, & Pedro-(eds), 2013).

6-4.4 - SECONDARY OUTPUT

6-4.4.1 - Stingless Bee as Pollinator

In using the inborne foraging affinity of a bee in the realm of crop production, the stingless bee is apt to be used in pollination services as a secondary output of meliponiculture (**Figure 372**).

6-4.5 - ISSUES IN POLLINATION SERVICES

6-4.5.1 - Mobile Beekeepers

Stingless beekeepers who transport long distances to provide pollination services tend to rely on artificial feeding, i.e., sugar water during transit. This artificial feeding also risks pests and disease spread (Heard, 1999).

6-4.5.2 - Artificial Queens

Many beekeepers resort to artificial queens to quickly multiply colonies to comply with pollination services demands. *In vitro* queen breeding in Apiculture has shown some degenerative effects on the genes required for the bee immune system. It is believed to have contributed to the decline of Apiculture worldwide. Further studies are needed on these possible negative effects before applying them to meliponiculture.

6-4.5.3 - Pesticides

Flowers sprayed with pesticides tend to keep the chemical residue in the pollen that may be foraged by bees in the next few days. The residual effects on bees show up as mutations in the newly hatched pupa that has been fed the contaminated pollen. A more immediate setback is the collateral eradication of Hemiptera (aphids, treehoppers and scale insects). Stingless bees rely heavily on honeydew provided by Hemiptera.

6-4.5.4 - Monoculture

A large acreage of Pineapple farming, Oil palm estates, Rubber plantations are examples of large-scale Monoculture. Other major cash crop concentration shows a need to employ commercial pollination services. However, in meliponiculture, foraging only one type of flowering vegetation for long periods has been detrimental to stingless bees.

Figure 398 Examples of fruit flowers visited by the stingless bee

6-4.6 - POLLINATION SERVICES

6-4.6.1 Arbitrary Predicaments

Many cases are yet to be resolved on pollination services when it comes to failures of the bee colonies during the service. Failures in the field may be due to many reasons (these precautions are usually taken in the meliponary but overlooked in customer farms). Pests and predators may be rampant in the farms. Other disputes include transportation costs of the beehives to the intended farm and back to the meliponary. The frequency and synchronization of pesticides in the customer farm are a big issue to be arbitrated. Agreeing to the appointment of an appropriate arbitrator is another issue.

6-4.6.1 - Profit Sharing

Portion allocation of improved yields versus pollination service charges is yet to be amicably standardized for both parties. Currently, the remunerations for the beekeepers are based on mutual agreements between the beehive owners and the customer farm. Some may agree on sharing improved yields (problems here in proving the actual yield improvement). Some beekeepers charge a fee for a set period of service, i.e., hive rental per period (problem here to ascertain the period for efficient pollination of different crops).

6-4.7 THREATS TO THE STINGLESS BEE NATURAL ECOSYSTEM

6-4.7.1 Habitat Destruction, Deforestation, Tree Felling

Tree felling in log hive acquisition as a cause of habitat destruction and deforestation, leading to one of the negative impacts on the whole ecosystem. Resource acquisition is an initial threat when enthusiasts in beekeeping look to acquire colonies for their meliponary. Tree felling is the first step to habitat destruction when they try to obtain a nest that has made its home in the hollow of a tree. Some are mature trees aged more than 100 years. The habitat destruction goes on as unscrupulous groups ravage the forests for their greed of attaining more and more stingless bee colonies. Growing on to the larger and larger scale inevitably leads to deforestation.

6-4.7.2 Intraspecific and Interspecific Competition

Intraspecific competition of a peculiar kind, namely attacks between colonies, may come about in the insect world due to overcrowding and subsequent aggressive swarming (Dollin, Dollin, & Rasmussen, 2015). As a consequence of habitat destruction, the remaining colonies drift and shift to smaller and smaller pockets of reliable vegetation (Jalil, 2014). This drift then creates new obstacles in resource partitioning and encroaches into each other's foraging territory. Unforced aggression results from the inadequacy of forage sources (Inoue, Salmah, Abbas, & Yusuf,1985).

6-4.7.3 Pesticides

The application of pesticides at forest fringes presents dangers in the natural habitat. Uncontrolled winds push airborne toxic chemicals deeper and deeper into wild shrubbery and sapling forests. Hive shifting from forest trees to village trees does not help the situation. Some species with pliable enough cerumen may practice 'entrance closing' or constriction, restricting the bee population foraging duties. The domino effect ends with the colony being weak with starvation (Jalil, 2018, 2019).

6-4.8 - THREATS IN THE DOMESTICATED STINGLESS BEE ECOSYSTEM

6-4.8.1 - Termitidae

A specialized termite group, with their gut having a high capacity to degrade lignocellulose. *Macrotermes bellicosus* is a species of *Macrotermes*. Its workers are about 0.14 in (3.6 mm), and soldiers are larger. *Bellicosus* means "combative" in Latin. The species is a member of a genus indigenous to Africa and South-East Asia. Control of termites is necessary to preserve the wooden log or hive box.

Macrotermes_bellicosus - (minor soldier)

Figure 399 Termites - Macrotermes bellicosus

The batumen shell of the stingless bee nest is a natural termite deterrent, and termites are never found inside the bee's nest. (Lumber cellulose is food for the termites). Meliponines can also combat the termites on a nest invasion at their nest entrances (Jalil, 2014, 2016). (**Figure 400**).

6-4.8.2 - Extreme Weather

Figure 400 Fungus Mycelium in nest boxes

Wet weather and high moisture content in the wood of the log hives and box hives can aggravate fungus growth (**Figure 401**). Fungus build-up may be inevitable in a Typhoon prone region; however, periodic maintenance can be done to avoid deterioration of colonies due to fungus (Jalil, 2018, 2019), (Jalil, & Roubik(ed), 2019). Early preventative measures can curb this. Lumber used for box making should be seasoned and completely dry (flame roasting may also help kill wood borer eggs).

A complete seal of the cavity is imperative. The box joints should be rebate jointed or tongue and groove jointed). Opening of hives (for whatever reason) should be done in an absolute dry environment. The box's insides intended for hive use can be coated with propolis extract to eradicate fungus spores (Jalil, 2014, 2016 & 2018). Protect the hive at all times from rain or wet winds.

6-4.9 - NATURAL RETARDANTS

This component includes predators, pests, enemies (in forage competition and resource acquisition), nest destroyers and parasitic inquilines (Michener, 1961) here. We look at mammals like bears with a penchant for honey (Rasmussen, 2008). The Malayan Sun Bear (*Helarctos malayanus*) and the Formosan Black bear *(Ursus thibetanus formosanus)* (Jalil, 2014, 2016 & 2018) (**Figure 402**)

"Glad to know about this tropical bear. I thought there were no native bears in tropical areas. The only bear that I know that lives closer to the tropics is the bear species in Taiwan. However, this country has a temperate, subtropical climate with native stingless bees. Though the Philippines has no native bear, we have native squirrels in Palawan. Not only in Palawan but we have also here in Laguna and Quezon", quoted by a Filipino beekeeper (Ciar, R. R., & et al. (n.d.)) (Cockerell, 1919).

6-4.9.1 - Sun Bears

Sun bears have been reported to destroy stingless bee nests in Malaysian forests. The sun bears are attracted to the bee's nests within reach. Claw marks have been sighted at tree trunks that host Meliponine hives.

There have not been reports of nests destroyed higher than the adult bears reach. For this reason, we find the smaller *Tetragonula* prefer building nests in tree branch cavities (Engel, Michener & Boontop, 2017), and *Lophotrigona canifrons* make their nests up to 15 meters above the ground (Jalil, 2016) (Schwarz, 1937 & 1939). A sun bear has raided a Meliponary in Central West Malaysia, leaving six devasted hives (Jalil, 2018).

Ursus thibetanus formosanus *Helarctos malayanus*

Figure 401 Sun Bears

6-4.9.2 - Squirrels

Squirrel attacks on logs have been reported in North Borneo and East Kalimantan. Predominantly, the culprit is the plantain squirrel, oriental squirrel or tricoloured squirrel (*Callosciurus notatus*) (**Figure 403**). The squirrels do not eat the bees but are probably attracted to the honey or the "bee-bread" (Sakagami, 1978). Its diet includes stingless bee larvae. Squirrels are not a very common predator, but reports have indicated occasional nest destruction. Even *Sundatrigona moorei* colonies in ant nests in the wild are always destroyed by squirrels.

Figure 402 Squirrel

6-4.9.3 - Ratel or Honey Badger

The Indian Ratel (*Mellivora capensis indica*) (Kerr, 1792) sometimes raids beehives in western Pakistan and western India (**Figure 404**). It also feeds on insects, frogs, tortoises, turtles, lizards, rodents, snakes, birds and eggs. In this category, we also include the Ratel or Honey Badger.

Figure 403 Indian ratel

6-4.9.4 - Pangolins

Figure 405 Pangolin

Pangolins are sometimes known as scaly anteaters. The name "pangolin" originates from the Malay word pengguling, meaning "one who rolls up", Now it is tenggiling in modern Malay, whereas in Indonesian, it is trenggiling; and in the Philippine languages, it is goling, tanggiling, or balintong (with the same meaning) Pangolins are an important regulator of termite populations in their natural habitats (**Figure 405 & 406**).

Figure 404 Left: Scratch marks of devastation evidence left behind by the Borneo Pangolin in Bahta, Central Kalimantan; Right https://commons.wikimedia.org/wiki/File:Pangolin_borneo.jpg

6-4.9.5 - Monkeys and Boars

Figure 406 Monkeys and Boars

This category shows nest destroyers usually are neither honey hunters nor insectivorous. They are attracted to the smell of sweetness and the congregation of the insects. The curiosity and intrigue urge them to displace and raid the beehive.

Monkeys here will include Macaques (*Macaca* spp., Monyet and Berok in Malay). Gibbons (*Hylobates lar*, Ungka in Malay)

Malaysian lar gibbon fur colouring varies from black and dark brown to light-brown, sandy colours. Selangor

silvered langur (*Trachypithecus selangorensis*, Lotong in Malay or Lutung in Java) (Figure 6-4.8).

Wild Boars (*Sus scrofa vittatus*, Babi Hutan in Malay) A small, short-faced and sparsely furred subspecies with a white band on the muzzle From Peninsular Malaysia and Indonesia from Sumatra and Java east to Komodo (**Figure 407**).

6-4.9.6 - Birds

The birds topping the list would be Bee-eaters (*Merops* spp, Berek-berek in Malay or Perek-perek in the Philippines), Magpie Robins (Murai in Malay), Flycatchers (*Terpsiphone* spp.), like Bee-eaters, are winter migrants to Southeast Asia, Rufous-collared Kingfisher (*Actenoides concretus*, Raja-udang in Malay) is found in South East Asia, including swallows and swiftlets plus many other numerous insectivorous birds (Jalil, 2014 & 2-16) (**Figure 408**).

Figure 407 Birds

6-4.9.7 - Bats

Figure 408 Bats

Fruit bats, especially *Eonycteris spelaea* (**Figure 409**), are the major pollinators of the durian, although the giant honey bees and stingless bees (meliponines) are frequent visitors to the flowers. Bats visited durian flowers at about 14 to 26 per inflorescence per night (Jalil, 2014 & 2016). They become enemies of stingless bees at dusk.

6-4.9.8 - Frogs and Lizards

Other enemies like ants, lizards, frogs, spiders and others are also noteworthy. These creatures will creep to the log hives if a "bridge" connects directly to a log/nest. Twigs or vegetative growth connecting to the logs, walls, poles, trees and the like make it easier to prey to the destroyer and logs/nest.

Various methods and stands (pedestal) may reasonably reduce (if not eliminate) the problem of the threat from below. For example, using steel pipes smeared with black oil can remove ants' threats, lizards, and frogs. Underlays may also be a bridge if it is not given the attention it deserves.

Positioning, orientation and protection of the hives to avoid predators and ensure survival in the Meliponary; a study (pers. obs.) of predators shows that the main culprits in a Malaysian Meliponary are frogs and lizards;

Ranidae: Typical Frogs, Rhacophoridae: Asian Tree Frogs, Microhylidae: Narrow-mouthed Frogs, Bufonidae True Toads (Baker, 2016).

Lizards include Chameleons (Sesumpah), (Mengkarung), Crested Lizards (Cicak Kobeng), Geckos (Tokek) and other small reptiles (**Figure 410**).

Figure 409 Frogs and Assorted Lizards

6-4.9.9 - Flies and Beetles

A few species of flies are identified as threats to bees as follows;

1. Diptera: Statiomyidae*:* *Hermetia illucens* (Black Soldier Fly)
2. Phoridae: *Dorniphora trigonae* (Phorid Flies)
3. Diptera: Syrphidae: *Cerial* cf. *ornatifrons* (Hover Fly, Flower Fly, Syrphid Fly)

The small hive beetle, *Aethina tumida*, introduced unintentionally from Africa, is a considerable threat to colony survival and performance. When the beetles access the nest of a weak colony, many eggs are quickly laid that hatch and consume the stores of both honey and bee bread. Larval infestation becomes so great that the colony may abscond (Jalil, 2016, 2018) (Heard, 2016) (**Figure 411**).

1. Coleoptera: Nitidulidae: *Aethina tumida* (Small Hive Beetle)
2. *Protaetia mandarinea* (Coleoptera: Scarabaeidae)
3. *Homalota trigonae* (Coleoptera: Staphylinidae)
4. Coleoptera: Nitidulidae: *Haptoncus luteolus* (Pollen Beetle)
5. *Brachypeplus auratus, B. basalis, B. meyricki, B. planus* (Coleoptera: Nitidulidae)

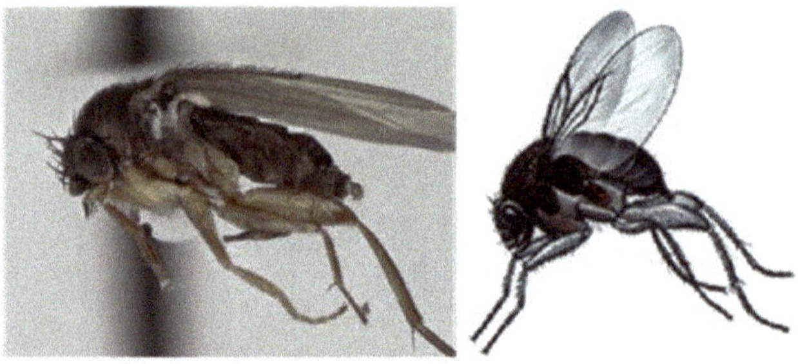

Phoridae: Dorniphora trigonae, Pseudohypocera and Melaloncha. (Phorid Flies)

Dorsal view of Phorid Fly at rest

Brachypeplus auratus, B. basalis, B. meyricki, B. planus (Coleoptera: Nitidulidae)

Aethina tumida (Small Hive Beetle)

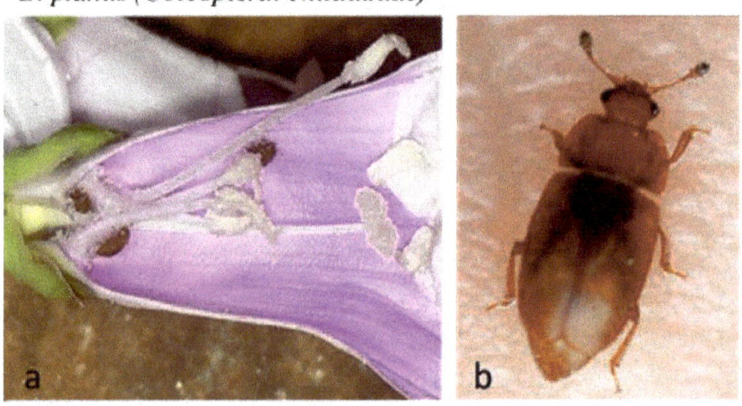

a. cross- section of morning glory corolla; b. & c. Pollen Beetle (Haptoncus luteolus)

Figure 410 Flies and Beetles

6-4.10 - CRAWLING ENEMIES

6-4.10.1 - Ants and Spiders

The most common species of ants and spiders are recognized as the threats are;

Formicidae: *Camponotus, Oeocophylla* and *Crematogaster* (Ants)

Thomasidae, Oxyopidae: (Spiders) (**Figure 412**)

Figure 411 Spiders and ants

6-4.10.2 - Praying Mantis and Hornets

Mantispidae is predatory carnivorous insects, which are indiscriminate ambush hunters. They regularly capture unsuspecting worker bees in their foraging runs. The mantis' favourite haunts are amongst lush vegetation and are a master of camouflage (**Figure 413**).

Hornets that preyed on honey bees in Fujian province, China, were identified as *Vespa mandarinia, V. bicolor, V. basalis, V. tropica baematodes, V. ducalis* and *V. affinis*. After witnessing that they also prey upon stingless bees, this information is considered, given an opportunity to forage honeydew from aphids. Experiments on control methods showed that the most effective was to smear captured hornets with poisonous dust ("Huichaoling dust") at the rate of 1 – 2 g per hornet.

Figure 412 Mantis and hornets

6-4.11 - OTHER INSECT ENEMIES

Wasps and hornets are very difficult to control, and their predation on bee colonies can be significant. Colonies of these insects are not permanent, and the damage to meliponines normally comes when the wasp workers have built up to maximum numbers. Where they are not endangered species, nests can be sought and eliminated. A range of screening and trapping techniques are used. At this time of the year, beekeepers may put out some sticky, sweet substance dissolved in a deepwater container to act as a trapping system. Elimination of landing boards may help. In Asia, the native *Apis cerana* has its defence mechanism; the guard bees "ball" the hornets killing them with their body heat.

6-4.12 - PESTS AND PARASITOIDS

6-4.12.1 - Mites and Moths

Among the mites and moths involved are;
Lepidoptera: Tineidae*:* *Achroia grisella* (Lesser Wax Moth)
Acaridae: *Eumellitiphis inouei* (Mycophagous Mites). Acari: Pyemotidae: *Pyemotes sp.* (Parasitic Mites) (**Figure 414**).

Figure 413 Moths and mites

6-4.12.2 - Megachilidae

In addition, Megachilidae such as *Megachile disjuncta, M. umbripennis* (Leaf-cutter bee) & other brood destroyers are also known among the threats (**Figure 415**).

Figure 414 Pests – Propolis robbing Megachilidae disjuncta and M. umbripennis.

6-4.13 - SNAKES AND RATS

All snakes are carnivores (**Figure 416**). Their diet depends on the species. Some eat warm-blooded prey (e.g., rodents, rabbits, birds), while others eat insects, amphibians (frogs or toads), eggs, other reptiles. Anything that the frog eats will be in its belly. But when the snake eats it, everything in the belly of the frog will be consumed as well as the frog itself.

Trimeresurus popeiorum -Pit Viper.

Figure 415 Snake

Rodents, especially rats and mice, can move into a weak colony or uncolonized hive, kill the bees with their nesting activity, or render an uncolonized hive uninhabitable. Keeping entrances small and carrying out regular inspections will prevent this from occurring. Sterilizing empty hives (this can be done over a fire) that have housed a rat or mouse family is necessary to eliminate the smell before the bees consider it attractive again.

6-4.14 - MELIPONICULTURE MANAGEMENT AND INDUSTRY IMPACT

6-4.14.1 – Countermeasures - Repellents and Larvicides

Many pests and inquilines have larvae that may destroy the meliponine brood (**Figure 417**); an ingenious manner of controlling these is applying larvicide at or before the pests' 3rd instar stage. The bee larva will not be affected because it is still within its brood cell. Larvicides include biological insecticides, such as microbial larvicides *Bacillus sphaericus* and *Bacillus thuringiensis israelensis*. Larvicides include other pesticides, such as temephos, methoprene, oils, and monomolecular films. I wouldn't use them…

Figure 416 Parasites and brood destroyers

6-4.14.2 - Owls and Feral Cats

Predators with ambush strategies like garden lizards and predating spiders will be kept at bay with the right type of predatory bird, like owls (**Figure 418**). As in traditional owl houses in paddy fields, owl houses can be incorporated to reduce bats, insectivorous birds, mice and frogs, thus giving the stingless bee a better chance of survival!

Figure 417 Domesticated owls

The bearcat (Binturong), also known as a feral cat, is carnivorous, and they eat chicken, birds, and snakes. They are Viverridae, where civet cats also belong, not a bear. Many beekeepers may ask a few frequent questions related to bearcat;

Q: Would they fall into the meliponiculture ecosystem?

A: Yes, because they are carnivores, they eat birds that eat stingless bees. "Simple philosophy on that basis", argues the beekeeper.

Q: Yes, in wild ecology, but meliponiculture domestication for the 3rd level food chain, we prefer a trainable predator of a bee predator. So, can you keep one and train it to catch the birds that eat your bees? Only then it can be organic control of predatory insectivorous birds.

A: You can train it if you catch a younger one.

The binturong (*Arctictis binturong*), also known as **bearcat**, is a viverrid native to South and Southeast Asia (**Figure 419**).

Figure 418 Bearcat or feral cat

6-4.14.3 - Meliponary Management

Meliponary management involves controlling the density of hives a farm may hold, i.e., having the appropriate number of hives in a particular farm area just enough to avoid overcrowding and balance vegetative forage and nest building material. Selection and placement of compatible intraspecies to avoid competition, provide optimum output results, and impact the whole scheme.

6-4.15 - POSITIVE IMPACT

6-4.15.1 - Medicinal and Wellbeing

Derivative products and value-added Propolis and Pot-pollen fill the medicinal pharmaceutical, nutraceutical and cosmeceutical bee products sector.

6-4.15.2 - Agronomic Yields

The yields of cash crops are greatly improved with Stingless bee pollination of agricultural plants. The yield improvements can range from 40% to 70%, depending on species compatibility and weather conditions.

6-4.15.3 - Health and Economy

Health and wealth ensue from the honey output, based mostly on sound and proper Meliponary Management. Appropriate harvesting and post-harvest techniques are critical in meliponiculture during the right balance of the floral, fruiting and weather seasons.

6-4.16 - ACKNOWLEDGEMENTS

I would like to thank the following people for their feedback and constructive criticism in the completion of this "Bubble": Rusphin, Ranau & Richard Tuaran, Sabah, Abdul Mushir KL, Malaysia; Albi Banjarmasin, Indonesia; Prof Muthuraman India; Ame Takashi Japan; JR Belmonte, Rizal, Luzon, Bernard Tuzon Anciado, Laguna, Luzon, Philippines.

6-4.17.1 - List of Wikimedia images

1. *Macrotermes bellicosus* (minor soldier)

https://commons.wikimedia.org/wiki/File:Macrotermes_bellicosus_minor_soldier.jpg

2. Ursus thibetanus formosanus

https://commons.wikimedia.org/wiki/File:Formosan_Black_Bear.JPG

3. Helarctos malayanus

https://commons.wikimedia.org/wiki/File:Sun-bear.jpg

4. *Callosciurus notatus*

https://commons.wikimedia.org/wiki/File:Squirrel--Bukit-Timah.jpg

5. *Mellivora c. indica*

https://commons.wikimedia.org/wiki/File:Indian_ratel.jpg

6. Philippine Pangolin

https://commons.wikimedia.org/wiki/File:Philippine_Pangolin_Curled-up_by_Gregg_Yan.jpg

7. Macaque Monkey

https://commons.wikimedia.org/wiki/File:Monkey_Kai_Bae.jpg

8. Silver langur

https://en.wikipedia.org/wiki/File:Silver_langur_with_baby.jpg

9. Lar gibbon

https://commons.wikimedia.org/wiki/File:Weisshandgibbon_tierpark_berlin.jpg

10. *Macaca nemastrina*

https://commons.wikimedia.org/wiki/File:Cercopithecidae_-_Macaca_nemastrina.jpg

11. *Sus scrofa vittatus*

https://commons.wikimedia.org/wiki/File:Banded_Pig_(Sus_scrofa_vittatus)_(8750051577).jpg

12. Oriental Magpie Robin

https://upload.wikimedia.org/wikipedia/commons/4/45/Male_Female_Oriental_Magpie_Robin_Photograph_By_Shantanu_Kuveskar.jpg

13. Green Bee Eater - *Merops* sp.

https://commons.wikimedia.org/wiki/File:Green_Bee_Eater_Pune.jpg

14. Amur Paradise Flycatcher -*Terpsiphone incei*

https://commons.wikimedia.org/wiki/File:Amur_Paradise_Flycatcher_(Terpsiphone_incei)_-_Lip_Kee_(3).jpg

15. Rufous-collared Kingfisher

https://commons.wikimedia.org/wiki/File:Rufous-collared_Kingfisher_-_Thailand_S4E3779.jpg

16. *Eonycteris spelaea*

https://commons.wikimedia.org/wiki/File:Eonycteris_spelea.png

17. *Trimeresurus popeiorum* -Pit Viper

https://en.wikipedia.org/wiki/File:Trimeresurus_popeiorum,_Pope%27s_pit_viper_-_Doi_Phu_Kha_National_Park_(46846412805).jpg

18. Binturong (*Arctictis binturong*)

https://commons.wikimedia.org/wiki/File:Binturong_in_Overloon.jpg

6-4.18 References to Chapter 6-4

Baker, N. (2016). *Amphibians of South East Asia*. Retrieved from Ecology Asia: http://www.ecologyasia.com/verts/amphibians.htm

Ciar, R. R., & et al. (n.d.). Foraging Behavior Of Stingless Bees (Tetragonula Biroi Friese): Distance, Direction And Height Of Preferred Food Source. *Uplb Bee Program.*

Cockerell, T. D. (1919). The Social Bees of The Philippines Islands. *Univeraity of Colarado.*

Dollin, A. E., Dollin, L. J., & Rasmussen, C. (2015). Australian And New Guinean Stingless Bees Of the Genus Austroplebeia Moure (Hymenoptera: Apidae)—a revision. *Zootaxa.*

Engel, M. S., Michener, C. D., & Boontop, Y. (2017). Notes on Southeast Asian Stingless Bees of the Genus Tetragonula (Hymenoptera: Apidae), with the Description of a New Species from Thailand. *American Museum Novitates.*

Heard, T. (1999). The Role Of Stingless Bees In Crop Pollination. *Annu. Rev. Entomol. 44:*, 183–206.

Heard, T. (2016). The Australian Native Bee Book, *ISBN 978-0-646-93997.*

Inoue, T., Salmah, S., Abbas, I., & Yusuf, E. (1985). 1985. Foraging Behavior of Individual Workers and Foraging Dynamics of Colonies of Three Sumatran Stingless bees. *Res. Popul. Ecol. 27:*, 373-392.

Jalil, A. (2014). *Beescape for Meliponines.* Singapore: Partridge Publishing.

Jalil, A. (2016). *Meliponine Identifier Pictoruial Guide.* Akademi Kelulut Malaysia.

Jalil, A. (2016). *World Meliponine Etymology of Taxonomic Nomenclature.* Akademi Kelulut Malaysia.

Jalil, A. H. (2018). *Southern Luzon Meliponiculture Tour.* Luzon, The Philippines: Akademu Kelulut Malaysia.

Jalil, A. H. (2019). *Visayas Meliponiculture Tour.* Cebu, The Philippines: Akademi Kelulut Malaysia.

Jalil, A. H., & Roubik(ed), D. W. (2019). *Filipino Meliponiculture and Beyond inc. MIMAROPA Tours.* Mindoro, The Philippines: Akademu Kelulut Malaysia.

Jalil, A., & Roubik(ed), D. (2016). *Handbook of Meliponiculture.* Akademi Kelulut Malaysia.

Jalil, A., & Roubik(ed), D. (2018). Handbook of Meliponiculture Vol 2. *Akademi Kelulut malaysia.*

Kerr, R., (1792). Class I. Mammalia: containing a complete Systematic Description, Arrangement, and Nomenclature, of all the known Species and Varieties of the Mammalia, or Animals which give suck to their young, (Feb) 1792: 27 UNNO. pp., [33]-400.

Michener, C. D. (1961). Observations on the Nests and Behavior of Trigona in Australia and New Guinea (Hymenoptera, Apidae). *American Museum Novitates.*

Rasmussen. (2008). Catalogue of the Indo-Malayan/Australasian stingless bees (Hymenoptera: Apidae: Meliponini).

Roubik, D. (2006). Stingless bee nesting biology. *Apidologie 37.*

Sakagami, S. F. (1978). Tetragonula Stingless Bees of the Continental Asia and Sri Lanka (Hymenoptera, Apidae). *Jour. Fac. Sci. Hokkaido Univ. Ser. VZ, 2001. 21(2).*

Schwarz, H. F. (1937). Bornean Stingless Bees Of The Genus Trigona. *Bulletin American Museum of Natural History.*

Schwarz, H. F. (1939). The Indo Malayan Species of Trigona. *Bulletin of AMNH.*

Thai, P. (2001). Foraging and pollination efficiency of the stingless bee, Tetragonisca angustula, on radish, Raphanus sativus, in a greenhouse, *M.Sc. thesis Utrecht University, The Netherlands.*

Vit, P., Roubik, D. W., & Pedro-(eds), S. M. (2013). Pot Honey - A Legacy of Stingless Bees. *Springer.*

Appendix A

Sap, latex or gum and resin are not the same, neither inconsistency nor viscosity. There are milk rubber or latex types; some use rubber, and some use sap. However, phloem sap is a sugary liquid, not a resin or gum, and is much the same as honeydew ejected from Hemiptera feeding on plant sap. Some mixed all the rubber, sap and rubber. Some do NOT use all that and only use DAMAR (In India, stingless bees are called damar bees) resin. Depending on the type of kelulut species. Never once did I type like that. Please accept that the next time I call it kelulut. You need to know the difference between plant secretions such as milk sap, latex, sap, rubber, gum and phloem. Resin, or in other words 'resin', is completely different. Ordinary trees do NOT have resin. ONLY resin trees have resin. Wax is not taken directly but in the digestion of the synthesis of nectar, phloem and other secretions from plants and secreted from certain glands in the body kelulut. Once you have this knowledge, what are the benefits, and how do you want to use it? I also want to know?

Abu Hassan Jalil: What is the difference between A. Latex and B. Sap?

A. Rubber liquid = latex ... milk from rubber trees/rubber/latex.

B. Sap exudate is sticky from the fruit's skin or like a milky white tree, e.g., young papaya fruit, young mango, jackfruit, cempedak, breadfruit etc. It's called milk rubber. So, milk rubber and latex are NOT the same.

Gum or Gum from the Gum tree species of eucalyptus and mangoes where the secretion is sticky like glue but not milky white.

The sap is a secretion from young trees or young wood 'Sapling'. There are also petals (flower petals), sap, sepal sap & rind (fruit skin) sap.

Phloem sap is a secretion from the shoots or sprouts of plants. Phloem and sap are the same things.

Resin from resin trees such as Keruing, Kapur, Meranti, Merawan, Cengal, Ulin, Balau etc. Also, Agathis oil resin tree sp. *Pinus* sp. *Araucaria* sp. *Thuja* sp. *Juniper* etc.

In Indonesia alone, there are various names and designations. For example, the names of bees without stings below?

Venacular Names of Stingless Bees throughout Indonesian Archipelago

SUMATRA

1. Linot (Aceh throughout)
2. Linot = Linoet (Aceh)
3. Gegeci (Aceh Tamiang)
4. Kekening (Aceh Gayo)
5. Sasau (Kerincim Sumatra)
6. Ben-ben/Keci-keci (Karo - N. Sumatera)
7. Guli Bombon (Tanjungbalai, Asahan, N.Sumatra)
8. Galo-galo (W. Sumatra, Kampar Riau, Palembang)
9. Gegele (Lahat - S. Sumatra)
10. Gala-gala (OKU S. Sumatra)
11. Gale-gale (OKI)
12. Gaalei (Bengkulu, Sumatra)
13. Likam/Until (W. Jawa & Lampung, S. Sumatra)
14. Likam/Ghibung (C. Lampung, S. Sumatra)
15. Thehen (E. Lampung, S. Sumatra)
16. Gegela/Kerebong (S. Lampung, S. Sumatra)
17. Puli/Pulut/Bon-bon (Mandailing Natal – N/ Sumatera)

BANGKA & BELITUNG

1. Madu Keran/Teran (Belitung)
2. Pelaket (Bangka Is.)

KALIMANTAN

1. Hamurep/Kalulut (Dayak of S. Kalimantan)
2. Udep (Dayak of E. Kalimantan)
3. Kelulut (Melayu, Dayak of W. Kalimantan, Riau, Kepulauan Riau*)
4. Kalulut (Banjar)
5. Kengkapit (G. thoracica) W. Kalimantan

JAVA & BALI

1. Teuweul (Sunda)
2. Teuwel-teuwel (Jakarta)
3. Klanceng (Madura, Tegal, Kediri, Magelang, Nganjuk, Jepara, Yogyakarta, Java throughout)
4. Lanceng – DIY(Jogjakarta)
5. Lilit (Kebumen)
6. Tiwuan (Tasikmalaya, Ciamis, Singaparna, C. Java)
7. Kele-kele/Kelo-kelo (Bali)
8. Aletan (Bali Nusa 2)
9. Kekele – Bali

LOMBOK & SUMBAWA

1. Nyanteng (East Lombok)
2. Nyiuk tales (C. Lombok)
3. Kledan (N. Lombok)
4. Teldan (N/ Lombok)
5. Kledan (N. Lombok) = Teldan
6. Nete (Bima, Sumbawa)

SULAWESI

1. Gala-gala (Mandar- W. Sulawesi)
2. Ketape/Kammu (C. Sulawesi)

3. Sarese (Poso – C. Sulawesi)
4. Merang (Plopo – S. Sulawesi)
5. Ummu (Watansoppeng – S. Sulawesi)
6. Emuuk/Emmu (Bugis)
7. Emmuk (Makasar, Sulawesi)
8. Pulo (Suku Tolaki, Kendari, Kolaka, Sulawesi)
9. Kaputi (Sorowako, S. Sulawesi)

MALUKU

1. Mae Mae Toher (Ambon Is., Maluku)
2. Kelkello (Saparua Is., Maluku)
3. Saseero (Seram Is., Maluku)
4. Bulpena (Haruku Is. Maluku)
5. Dodahala (N. Halmahera, Maluku)

PAPUA

1. Ndokof (Marori tribe)
2. Mentya (Tetrogonula mellipes) and
3. Bress (Austroplebeia cincta) Kanum tribe.
4. Bebiyem (Biak language, West Papua)

MALAYSIA (KadazanDusun)

1. Tantadan (Sabah, N. Borneo in general *Heterotrigona itama*)
2. Tantadan Bohuang Oitom (Bohuang (bear), Oitom (black). for *Geniotrigona thoracica*
3. Tantadan Bohuang Osilou. Osilou (yellow) for *Lepidotrigona terminata*
4. Tantadan Osilou/Gonsolod. Gonsolod (intruder).

BRUNEI

1. Kalimuduk (Brunei) *
2. Antadan = *G. thoracica,*
3. Kelulut Duung = *T. fuscobalteata*

A comparison with vernacular names of Stingless Bees in The Philippines.

1.1Kabuyay – clustered brood in Romblon/Sibuyan Island- Sibuyanon and Romblomanon

2. Kadulet - among locals of Macarascas in Palawan.

4. Kiwət - Waray of Northern Samar.

5. Kiwot - the Visayan term for stingless bees among beekeepers

6. Kiyut - In Davao, the term refers to both clustered and layered brood

7. Kiyut - probably the cute Kiwot of Oriental Mindoro

8. Kiyut – or kiwot for Tablas Island clustered brood

9. Lap-og - For layered brood in Sibuyan Island

10. Lib-og – layered brood in Tablas Island

Li-bog - refers to stingless bees in Panay Island. propolis as "kabulay" in Panay but Aklan its kabuyay for Aklanon, letter L is Y.

12. Lukot - Marinduque dialect, some there also call it kiwot or kiyot

Lukot –Tagalog term for stingless bees

14. Lukutan - or Lulukot here in Sison, Pangasinan Luzon.

15. Mahsam - possibly due to sour SB honey from Iraya Mangyan of Paluan

16. Mamulit - among Tagbanua of Buenavista in Palawan

17. Monggo Monggo – Clustered brood SB in Luzon

18. Mugo Mugo – refers to SB in Guimaras

19. Pukyutan - stingless bee is called pukyutan in Kalayaan, Laguna

20. Quiot - in Isabel, Leyte, Visayas it's called "quiot".

21. Takpal - layered brood in Negros Occidental

22. Tamusan or Laholoy - among the matigsalo Manobo tribe of Bukidnon

Tegteg - stingless bees for Ilocanos in Ilocos Province.

24. Tutoldan - the name of stingless bees in Cuyonon language of Cuyo Island.

Appendix B

List of wood in Indonesia

from Indonesian Wikipedia, the free encyclopedia

Indonesia has around 4,000 tree species, which can be used as building wood. But to date, only around 400 species (10%) have economic value and even fewer, 260 species, classified as trade timber. [1] The following is a list of the names of wood or timber groups according to their trade names, following the Attachment to the Decree of the Minister of Forestry Number: 163 / Kpts-II / 2003 dated May 26, 2003, concerning the Grouping of Timber Types for the Imposition of Forestry Fees; with some adjustments.

Meranti (Shorea) Type Group / Commercial Group One

No.	**Trade Name**	**Botanical Name**	**Vernacular Name**
1.	Agatis	*Agathis* spp.	Damar (Jw.), dama (Slw.), damar bindang (Klm.), damar sigi (Smt.). (Ingg.): *kauri pine*.
2.	Balau	*Shorea* spp. (e.g., *S. materialis* Ridl., *S. maxwelliana*King, *S. scrobiculata* Burck); *Parashorea* spp.	Damar laut (Smt.), semantok (Aceh), amperok, anggelam, selangan batu (Klm.)
3.	Balau Merah	*Shorea* spp. (e.g., *S. collina* Ridl., *S. guiso* (Blanco) Bl.)	Balau laut, damar laut merah, batu tuyang, putang, lempung abang. Ingg.: *red selangan*.
4.	Bangkirai	*Shorea* spp. (e.g., *S. kunstleri* King, *S. laevis* Ridley, *S. laevifolia* Endert); *Hopea* spp. (e.g. *H. celebica* Burck, *H. semicuneata* Sym.)	Benuas, balau mata kucing, hulo dereh, puguh, jangkang putih, kerangan (Smt.), bubuh (Bk.)
5.	Damar	*Araucaria* spp. (e.g., *A. cunninghamii* D. Don, *A. hunsteinii*K.Schum.)	Alloa, ningwik, pien (Pap.). Ingg.: *araucaria*.
6.	Durian	*Durio* spp. (terutama *Durio carinatus* Mast.); *Coelostegia* spp.	Durian burung, lahong, layung, apun, begurah,

			punggai, durian hantu, enggang
7.	Gia	*Homalium tomentosum* (Roxb.) Benth., *Homalium foetidum*(Roxb.) Benth.	Delingsem (Jw.), kayu batu, melunas, kayu kerbau, momala (Slw.)
8.	Giam[2]	*Cotylelobium* spp. (e.g., *C. burckii* Heim, *C. lanceolatum* Craib, *C. melanoxylon* Pierre	Giam durian, resak bukit tembaga; giam padi, resak daun kecil, resak batu; giam tembaga, resak daun lebar; resak gunung
9.	Jelutong	*Dyera* spp.	Pulai nasi, pantung gunung, melabuai
10.	Kapur	*Dryobalanops* spp. (di antaranya *D. oblongifolia* Dyer, *D. sumatrensis* (Gmelin) Kosterm.)	Kamper (kayu), kayu kayatan, empedu, keladan
11.	Kapur Petanang	*Dryobalanops oblongifolia* Dyer	Kapur guras (Smt.), kapur paya (Mly.), kelansau (Swk.)
12.	Kenari	*Canarium* spp., *Dacryodes* spp., *Santiria* spp., *Trioma* s pp.	Kerantai, ki tuwak, binjau, asam-asam, kedondong (kedundung), resung, bayung, ranggorai, mertukul
13.	Keruing	*Dipterocarpus* spp. (e.g., *D. applanatus* V.Sl., *D. baudii* Korth., *D. elongatus* Korth. etc.)	Keruing arong, kekalup; Lagan sanduk, mara keluang; Keruing tempudau; tempurau, merkurang, kawang, apitong
14.	Kulim	*Scorodocarpus borneensis* Becc.	Kayu bawang hutan (Klm.)
15.	Malapari	*Pongamia pinnata* (L.) Pierre	Malapari
16.	Matoa	*Pometia* spp.; e.g., *P. pinnata* Forster & Forster, *P. ridleyi* King	Kasai, taun, kungki, hatobu, kayu sapi (Jw.), tawan (Mlku.), ihi mendek (Irian Jaya)

17.	Medang	*Cinnamomum* spp.	Sintuk, sintok lancing, ki teja, ki tuha, ki sereh, selasihan
18.	Meranti Kuning	*Shorea* spp. (di antaranya: *S. acuminatissima* Sym., *S. balanocarpoides* Sym., *S. faguetiana* Heim, *S. gibbosa*Brandis, *Shorea scollaris* V.Sl.;	Damar hitam, damar kalepek; Damar hitam katup; Bangkirai guruk, karamuku; Damar buah, mereng-kuyung; Damar tanduk. Ingg.: *yellow seraya.*
19.	Meranti Merah	*Shorea* spp. (di antaranya: *S. johorensis* Foxw., *S. lepidota*BI., *S. leprosula* Miq., *S. ovalis* BI., *S. palembanica* Miq., *S. platyclados* V.Sl. ex Foxw., *S. leptoclados* Sym., etc.)	Majau, meranti merkuyung; Meranti ketrahan; Meranti tembaga, kontoi bayor; Meranti kelungkung; Tengkawang majau; Banio, ketir; Seraya merah, campaga, lempong, kumbang, meranti ketuko, cupang. Ingg.: *red seraya, red lauan.*
20.	Meranti Putih	*Shorea* spp. (e.g., *S. assamica* Dyer, *S. bracteolata*Dyer, *S. javanica* K. et. Val., *S. lamellata* Foxw., *S. ochracea*Sym., *S. retinodes* V.SI., *S. virescens* Parijs, *S. koordersi*Brandis, etc.)	Damar mesegar; Bunyau, damar kedontang; Damar mata kucing, damar kaca, damar kucing; Damar tunam, damar pakit; Damar kebaong, baong, bayong, baung, belobungo, kontoi tembaga; Balamsarai, damar mansarai; Damar maja, kontoi sabang; Kikir, udang, udang ulang, damar hutan, anggelam tikus, maharam

			potong, pongin, awan punuk, mehing (Smt., Kal.); Damar lari-lari, lalari, temungku, tambia putih (Slw.), Damar tenang putih, hili, honi (Mlku.). Ingg.: *white meranti*.
21.	Merawan	*Hopea* spp. (e.g., *H. dasyrrachis* V.Sl., *H. dyeri* Heim, *H. sangal* Korth., etc.)	Tekam, tekam rayap; Bangkirai tanduk, emang, amang besi; Cengal, merawan telor; Ngerawan, cengal balau
22.	Merbau	*Intsia* spp. (terutama *I. bijuga* O.K., *I. palembanica* Miq.)	Merbau asam, ipi (NT.), kayu besi (Papua); Ipil, anglai, maharan; Tanduk (Mlku.)
23.	Mersawa	*Anisoptera* spp. (e.g., *A. laevis* Ridl., *A. marginata* Korth., *A. thurifera* Bl.)	Cengal padi, damar kunyit; Masegar (Smt.), ketimpun (Klm.), mersawa daun besar; tabok, tahan
24.	Nyatoh	*Palaquium* spp., *Payena* spp., *Madhuca* spp.	Suntai, balam, jongkong, hangkang, katingan, mayang batu, bunut, kedang, bakalaung, ketiau, jengkot, kolan
25.	Palapi	*Heritiera* (*Tarrietia*) spp.; e.g., *H. javanica* (Bl.) Kosterm., *H. simplicifolia* (Mast.) Kosterm., *H. littoralis* Ait., *H. sylvatica* S. Vidal	Mengkulang, teraling; Dungun, talutung, lesi-lesi.
26.	Penjalin	*Celtis* spp.	Rempelas, ki jeungkil, ki endog (Sd.), cengkek (Jw.), pusu (Sumbawa)

27.	Perupuk	*Lophopetalum* spp.; e.g., *L. javanicum* (Zoll.) Turcz., *L. multinervium* Ridl., *L. subobovatum* King, *L. wightianum* Arn.	Kerupuk (Smt.), pasana (Klm.), mandalaksa (Jw.), aras
28.	Pinang	*Pentace* spp.	Melunak, ki sigeung, ki sinduk, kelembing
29.	Pulai	*Alstonia* spp. (e.g. *A. pneumatophora* Back., *A. scholaris* R.Br., *A. spatulata* Bl., *A. macrophylla* Wall., *A. spectabilis* R.Br.)	Kayu gabus, rita, gitoh, bintau, basung, pule, pulai miang. Ingg.: *white cheesewood, milkwood, milk y pine*.
30.	Rasamal a	*Altingia excelsa* Noroña	Tulasan (Smt.), mandung (Min.), mala (Jw.)
31.	Resak	*Vatica* spp.; e.g., *V. maingayi* Dyer, *V. oblongifolia* Hook.f., *V. rassak* Bl.	Damar along, resak putih

Mixed Jungle Type / Commercial Group Two

No.	Trade Name	Botanical Name	Vernacular Name
1.	Bakau	*Rhizophora* spp. dan Bruguiera spp	Tumu, Lenggadai, Jangkar, Tanjang, Putut, Busing, Mata buaya
2.	Bayur		Balang, Walang, Wadang, Wayu
3.	Benuang	*Octomeles sumatrana* Miq.	Benuang bini (Klm.), winuang (Slw.)
4.	Berumbung	*Adina minutiflora* Val.); *Pertusadina* spp.	Kayu lobang, Barumbung, Kayu gatal
5.	Bintangur	*Calophyllum* spp.; e.g., *C. calaba* L., *C. inophyllum* L., *C. papuanum* Lauterb., *C. pulcherrimum* Wall.ex Choisy, *C. soulattri* Burm.f.	Bintangor, penaga; Nyamplung; Sula

			tri; Bunoh, bintangur bunut
6.	Bipa	*Pterygota* spp.	Kayu wipa
7.	Bowoi	*Serianthes minahassae* Merr. & Perry (Syn. *Albizia minahasae*Koord.)	Rayango, Merang, Terangkuse
8.	Bugis	*Koordersiodendron pinnatum* Merr.	Grepau
9.	Cenge	*Mastixia rostrata* BI.	Cenge, Cingo
10.	Duabanga	*Duabanga moluccana* BI.	Benuang laki, Takir, Aras, Raju mas
11.	Ekaliptus	*Eucalyptus* spp.; e.g., *E. alba* Reinw.ex Bl., *E. deglupta* Bl., *E. urophylla* S.T. Blake	Kayu putih; Leda, aren (Mlku.), tampai; Ampupu (Timor),
12.	Gelam	*Melaleuca* spp.	Kayu Putih
13.	Gempol	*Nauclea* spp.	Wosen, Klepu pasir, Anggrit
14.	Gopasa	*Vitex* spp.	Teraut, Laban
15.	Gerunggang/Derum	*Cratoxylum* spp.; e.g. *C. arborescens* (Vahl) Bl., *C. cochinchinense* (Lour.) Bl.	Madang baro; Mampat, butun; kemutul, temau; edat
16.	Jabon	*Anthocephalus* spp. (*A. chinensis* (Lamk.) A.Rich ex Walp. dan *A. macrophyllus* (Roxb.) Havil.)	Kelampayan (Mly.), laran (Klm.), semama (Amb.). Ingg.: *cadamba*.
17.	Jambu-jambu	*Syzygium* spp.	Kelat, Ki tembaga, Jambu
18.	Kapas-kapasan	*Exbucklandia populnea* R. Brown	Hapas-hapas, Tapa-tapa, Leman
19.	Kayu kereta	*Swintonia* spp.	Rengas sumpung, Merpauh, Bagel mirah
20.	Kecapi	*Sandoricum* spp.	Papung, Kelam, Sentul
21.	Kedondong Hutan	*Spondias* spp.	Coco, Kacemcem leuweung

22.	Kelumpang	*Sterculia* spp.	Kepuh, Kalupat, Lomes
23.	Kembang Semangkok	*Scaphium macropodum* J. B.	Kepayang, merpayang (Smt.)
24.	Kempas	*Koompassia malaccensis* Maing.	Hampas, impas, tualang ayam
25.	Kenanga	*Cananga* sp.	Kananga
26.	Keranji	*Dialium* spp.; e.g., *D. indum* L., *D. platysepalum* Baker, *D. procerum* (v.Steen.) Stey	Kayu lilin; Maranji
27.	Ketapang	*Terminalia* spp.	Kalumpit, Klumprit, Jelawai, Jaha
28.	Ketimunan	*Timonius* spp.	Seranai, Temirit, Kayu reen
29.	Lancat	*Mastixiodendron* spp.	Kundur, Modjiu, Raimagago
30.	Lara	*Metrosideros* spp. dan *Xanthostemon* spp.	Lompopaito, Nani, Langera
31.	Mahang	*Macaranga* spp.	Merkubung, Mara, Benua
32.	Medang	*Litsea firma* Hook f.; *Dehaasia* spp.	Manggah, Huru kacang, Keleban, Wuru, Kunyit
33.	Mempisang	*Mezzetia parviflora* Becc.; *Xylopia* spp.; *Alphonsea* spp.;*Kandelia candel* Druce	Mahabai, Hakai rawang, Empunyit, Jangkang, Banitan, Pisang-pisang
34.	Mendarahan	*Myristica* spp., *Knema* spp.	Darah-darah, Tangkalak, Au-au, Ki mokla, Kumpang, Kayu luo, Huru
35.	Menjalin	*Xanthophyllum* spp.	Lilin, Ki endog, Segi landak
36.	Mentibu	*Dactylocladus stenostachys* Oliv.	Jongkong, merebung
37.	Merambung	*Vernonia arborea* Han.	Merambung, sembung

38.	Punak	*Tetramerista glabra* Miq.	Kayu malaka (Smt.), cerega (Klm.)
39.	Puspa	*Schima* spp.; terutama *S. wallichii* Korth.	Seru (Jw.), simartolu (Smt.), madang gatal (Klm.)
40.	Rengas	*Gluta aptera* (King) Ding Hou	Rengas tembaga, Rangas
41.	Saninten	*Castanopsis argentea* A. DC.	Sarangan (Jw.), ki hiur (Sd.), kalimorot
42.	Sengon	*Paraserianthes falcataria* (L) Nielsen	Jeungjing, Tawa kase, Sika (Maluku)
43.	Sepat	*Berrya cordofolia* Roxb.	Waru gunung, Kalong
44.	Sesendok	*Endospermum* spp.; e.g., *E. diadenum* (Miq.) Airy Shaw, *E. moluccanum* (T & B) Kurz, *E. peltatum* Merr.	Sendok-sendok, kayu labuh (Smt.), kayu bulan (Mly.), garung (Klm.); Kayu raja (Mlku.)
45.	Simpur	*Dillenia* spp.; e.g., *D. grandifolia* Wall., *D. obovata* Hoogl., *D. pentagyna* Roxb.	Sempur, segel, janti, dongi
46.	Surian	*Toona sureni* Merr.	Suren, kalantas
47.	Tembesu	*Fagraea* spp.; e.g., *F. fragrans* Roxb., *F. sororia* J.J. Sm.	Tomasu (Smt.), kulaki (Slw.), malbira, ki tandu
48.	Tempinis	*Sloetia elongata* Kds.	Damuli, Kayu besi
49.	Tepis	*Polyalthia glauca* Boerl.	Banitan, Pemelesian, Kayu tinyang, Kayu bulan, Banet, Kayu kalet
50.	Tenggayun	*Parartocarpus* spp.	Buku ongko, Pejatai, Purut bulu
51.	Terap	*Artocarpus* spp.	Cempedak, Kulur, Tara, Teureup
52.	Terentang	*Campnosperma* spp.; e.g., *C. auriculatum* (Bl.) Hook.f., *C. brevipetiolatum* Volkens, etc.	Tumbus (Smt.), pauh lebi

53.	Terentang Ayam	*Buchanania* spp.	Pauhan, Antumbus, Talantang
54.	Tusam	*Pinus* spp.	Pinus, Damar batu, Uyam
55.	Utup	*Aromadendron* sp.	Utup

Ebony Timber / Precious Group One

No.	Trade Name	Botanical Name	Vernacular Name
1.	Eboni bergaris	*Diospyros celebica* Bakh.	Maitong, Kayu lotong, Sora, Amara
2.	Eboni Hitam	*Diospyros rumphii* Bakh.	Kayu hitam, Maitem, Kayu waled
3.	Eboni	*Diospyros* spp.; e.g. *D. areolata* King et G., *D. cauliflora* BI., *D. ebenum* Koen, *D. ferrea* Bakh., *D. lolin* Bakh., *D. macrophylla* BI.	Baniak, Toli-toli, Kayu arang, Kanara, Gito-gito, Bengkoal, Malam

Fine Timber / Precious Group Two

No.	Trade Name	Botanical Name	Vernacular Name
1.	Bongin	*Irvingia malayana* Oliv.	Pauh kijang, Sepah, Kayu batu
2.	Bungur	*Lagerstroemia speciosa* Pers.	Ketangi, wungu (Jw.), tekuyung, benger
3.	Cempaka	*Michelia* spp., *Elmerrillia* spp.	Minjaran, Wasian, Manglid, Sitekwok, Kantil (Jw.), Capuka
4.	Cendana	*Santalum album* L.	Kayu kuning, Lemo daru

5.	Dahu	*Dracontomelon* spp.; e.g., *D. dao* Merr. & Rolfe, *D. mangiferum* Bl.	Dao, basuong (Smt.), sengkuang (Mly.), koili
6.	Johar	*Senna* spp.	Juar, Trengguli, Sebusuk, Bobondelan
7.	Kuku	*Pericopsis mooniana* Thw.	Kayu laut, Papus, Nani laut
8.	Kupang	*Ormosia* spp.	Kayu ruan, Saga
9.	Lasi	*Adina fagifolia* Ridl.	Adina, Kilaki
10.	Mahoni	*Swietenia* spp.; e.g., *S. macrophylla* King, *S. mahagoni* (L.) Jacq.	Mahoni
11.	Melur	*Dacrydium* spp.; *Podocarpus* spp. dan *Phyllocladus* spp. E.g. *Dacrydium junghuhnii* Miq.	Alau, cemantan (Klm.); Jamuju, kayu embun (Slw.), sampinur bunga (Smt.);Sampinur tali; Kayu cina; Ki merah, Sandu
12.	Membacang	*Mangifera* spp.	Ambacang, Asam, Limus piit, Mempelam, Wani, Mangga
13.	Mindi	*Melia* spp.; terutama *M. azedarach* L.	Bawang kungut
14.	Nyirih	*Xylocarpus granatum* J. Konig	Nyireh, Niri
15.	Pasang	*Quercus* spp.	Mempening, Baturua, Kasunu, Triti
16.	Perepat darat	*Combretocarpus rotundatus* Dans.	Marapat, Teruntum batu
17.	Raja Bunga	*Adenanthera* spp.	Saga, Segawe, Klenderi

18.	Rengas	*Gluta* spp.; *Melanorrhoea* spp.	Ingas, Suloh, Rangas, Rengas burung
19.	Ramin	*Gonystylus bancanus* Kurz	Gaharu buaya, Medang keladi, Keladi, Miang
20.	Sawo kecik	*Manilkara* spp.; e.g., *M. fascicularis* H.J. Lam & Maas Geest., *M. kauki* (L.) Dub.	Subo, Ki sawo
21.	Salimuli	*Cordia* spp.	Kendal, Klimasada, Purnamasada
22.	Sindur	*Sindora* spp.; e.g. *S. bruggemanii* de Wit, *S. coriacea* Maing., *S. wallichii* Graham	Sepetir (Mly.), sasundur (Klm.), mobingo (Slw.)
23.	Sonokembang	*Pterocarpus indicus* Willd.	Angsana, Linggua, Nala, Candana
24.	Sonokeling	*Dalbergia latifolia* Roxb.	Linggota, sono sungu, sonobrits
25.	Sungkai	*Peronema canescens* Jack	Jati seberang, Jati londo
26.	Tanjung	*Mimusops elengi* L.	Sawo manuk (Jw.), karikis (Slw.)
27.	Tapos	*Elateriospermum tapos* BI.	Kelampai, Setan, Kedui, Wayang
28.	Tinjau belukar	*Pteleocarpus lampongus* Bakh.	Lontar Kuning
29.	Torem	*Manilkara kanosiensis* H.j. L. et B. M.	Sawai, Torem
30.	Trembesi	*Samanea saman* Merr.	Ki hujan
31.	Ulin	*Eusideroxylon zwageri* T.et B.	Kayu besi, bulian, kokon
32.	Weru	*Albizia procera* Benth.	Beru, Ki hiyang, Bengkal

References -Appendix B

1. ^ Soerianegara, I. and RHMJ. Lemmens (eds.). 2002. Southeast Asian Vegetable Resources 5 (1): Major commercial timber producing trees. PROSEA - Library. Jakarta. ISBN 979-666-308-2. Hal. 7

Main Index

Abu Hassan Jalil

Began Bee culture career as an Apiary Landscaper as early as 1983, Meliponary Landscaping since 2010. Actively involved in Indo-Malayan Meliponine Conservation. Cooperative Director at Koperasi Meliponi KL Bhd 2013 to present. Beescape Consultant and Meliponine Species Behavioral Scrutiny at Malaysian Genome Institute Meliponine Repository and Consultant to numerous Stingless Bee Farms in Malaysia and Brunei. Principal at Akademi Kelulut Malaysia Sdn Bhd (Malaysian Academy of Meliponiculture).

www.ingramcontent.com/pod-product-compliance
Lightning Source LLC
LaVergne TN
LVHW071200210926
848482LV00014B/695

* 9 7 8 1 9 1 3 8 1 1 1 0 5 *